Baby Universe

Vol. II

From Big Bang to Big Crunch!

Author

Mr. Scientific

In the loving memory of my father,

Rajendra Singh

Contents

Introduction

Modern humans first walked the Earth about 200,000 years ago. Two hundred thousand years may sound like a long time, but it is minuscule compared to the age of this vast cosmic arena. By the time we opened our eyes and began looking up at the night sky, most of the things we observe today had already happened. We might never be able to understand some of them, but still, there is a lot of the universe just waiting to be discovered.

In the search for answers, we have done things that no other species on this planet is capable of doing. We have launched telescopes millions of miles into space, split the atom, and cracked the DNA code. We have even created our own gods— gods that give us a sense of belonging, a purpose to live and die for, and a 'universe' that makes sense.

Our understanding of the universe has gone through drastic changes throughout human history. We have come all the way from knowing nothing about the universe to understanding just how little we actually know about it. From a geocentric model in which Earth is the center of the universe and humans are the special beings of God, to the Big Bang model where we are not so special. We have also learned that there could be thousands of other civilizations like ours, or even more advanced ones, roaming the universe.

We have built many powerful telescopes such as Hubble, James Webb, and the Very Large Array to understand the outer universe. However, to understand the inner universe, we have developed powerful particle accelerators, such as the Large Hadron Collider and the Relativistic Heavy Ion Collider. The study of the deep universe is the study of our universe's past. Suppose we are observing a star one billion light-years away. We are not seeing it as it is today; instead, we are seeing it as it was one billion years ago. Suppose there is an alien civilization sixty-six million light-years away with advanced technology to see our planet. They would see dinosaurs wandering the Earth—or even a giant asteroid wiping out most of the life forms on this planet—if they could get their timing just right.

So, let me take you on this journey. We will start from before the beginning of the universe, sail through newborn stars and galaxies, understand the complex black holes, investigate the possibility of advanced alien civilizations, and explore how our universe could end.

I am sure this book will make you fall in love with the universe!

Timeline of History

Event	Number of Years Ago
The Big Bang	13.8 billion
Formation of Early Atoms	13.8 billion (380,000 years after the Big Bang)
Early Stars and Galaxies	13.6 billion
Formation of Sun	4.6 billion
Formation of Earth	4.54 billion
Formation of Moon	4.51 billion
Oldest Rocks Found	4 billion
The Beginning of Life	3.5 billion
Oxygenation of Earth	2.4 billion
Complex Cells	2 billion
Multicellular Life	800 million

Life Moved from Ocean to Land	530 million
Vertebrates (Creatures with Backbone)	525 million
Fish	500 million
Land Plants	450 million
Insects and Seeds	400 million
Reptiles	300 million
Dinosaurs	230 million
Mammals	200 million
Flowers	130 million
Mammals	225 million
Dinosaur Extinction	65 million
Primates	60 million
Apes	15 million

Human-Like Creatures	4 million
Primitive Humans	2.5 million
Use of Stone Tools	2.5 million
Domestication of Fire	400,000
Modern Humans (Homo-Sapiens)	200,000
End of the Ice Age	12,000
Agriculture	10,000
Development of Wheels	6,500
Iron Age	3,000
Invention of Telescope	400
Moon Landing	Year 1969*
First Black Hole Image	Year 2019*

PART-I

The Origin of the Universe

Before the Big Bang

A lot of people ask me: "We know about the Big Bang explosion and how it created the universe, but what existed before the Big Bang? What was there when there was no universe?" I often answer with a simple question: "Do you know who or what you were before you were born?" The answer is obvious: we do not know. All that we know comes from what we have learned throughout our lives, and the universe is no different.

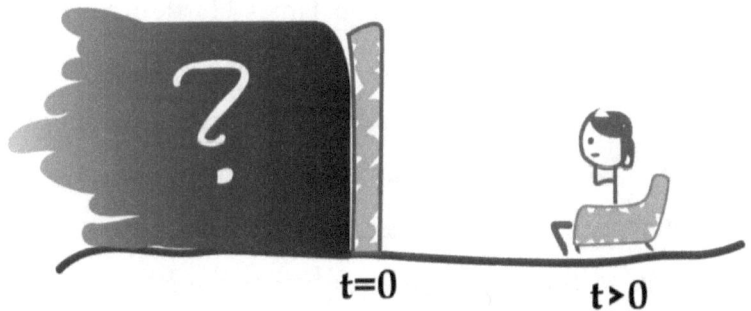

13.8 billion years ago, our entire universe existed in a tiny singularity. After the Big Bang explosion occurred, space and time came into existence. When there was no universe, there was no space, and there was no time. The terms 'before' and 'after' are strictly bound by the laws of time; that is how we use them. But since there was no time, there was no 'before' either. Time itself was nonexistent before the explosion and began counting as the Big Bang occurred. So, when we ask ourselves "what existed before the Big Bang?", we are asking the wrong question.

Throughout our lives, we have grown up knowing there is a past. We know what we did yesterday and the day before that. The human brain is trained in a way that makes it hard for us to imagine a universe without a past—just like it is hard for you to imagine who or what you were before you were born.

Hawking-Hartle Model

Physicists Stephen Hawking and James Hartle have tried to answer this question through the Hawking-Hartle model. This model shows that it is possible to create a model of the universe in which asking what happened before the Big Bang is the same as asking, "Where is the north of the North Pole?" According to them, once you go before the Big Bang, time does not exist, and the universe becomes pointless in the absence of time. Time, as we know it, is just one of the universe's properties, and we still have a lot to learn about the nature of time. Our understanding of time is limited to how we measure it using our clocks. The universe itself is indeed bound to follow the laws of time. However, it becomes timeless when we go beyond its beginning.

Before we knew about genetics and started sequencing DNA, philosopher John Locke advocated for the Blank Slate theory. This theory argues that when a child is born, they have no built-in traits. They do not have any beliefs, desires, memories, thoughts, intentions, or a wish of their own. They are like a blank slate. How children study in their early years, what they learn, and the environment they grow up in determine their personality. In general, this theory seems to hold some truth. It is common sense that you need to study astronomy and not animal behavior to become a good astronomer. However, the reality can be a little astonishing.

Has it ever happened that you went to see a doctor and they asked you if anyone in your family has a particular disease? It turns out that there are various traits that we inherit from our family. Different physical characteristics, such as bone structure, the color of your eyes, hair, and how tall you are, are inherited from your family. Also, certain diseases run in some families. There is no denying that education plays an essential role in who you are, but where you come from also makes a significant part of it.

Why are we talking about biology? Like the Blank Slate theory, some theories gave us an early model of the universe. So, before we jump to the Big Bang—the current leading model of the universe in cosmology—we must talk about one of its most renowned historical rivals: the Steady-State Theory.

THE STEADY-STATE THEORY

Sir James Jeans first put forward the Steady-State theory in the 1920s. As time passed, more observations came in, and the theory was revised by Hermann Bondi, Thomas Gold, and Fred Hoyle. The Steady-State theory argues that the universe is constantly expanding. As it expands, it creates more and more matter on its own. This new matter forms new stars and galaxies at the same rate at which the older ones become unobservable due to the expansion of the universe. As a result, the mean density of matter throughout the universe remains the same. So, it doesn't matter in which direction you are looking or what your observation point is; the universe will always look the same.

The Steady-State theory further argues that we live in a universe without a beginning or an end. The universe has existed forever, and it will exist infinitely. This theory produces a map of the universe that is infinite, with no beginning or end, and does not change with time. The Steady-State theory was famous in the 1950s, but today, most scientists do not accept it—especially after we started observing the deeper universe. First of all, no new matter is being created from a thin vacuum. All the new stars and planets born in the universe are formed from matter created in the Big Bang—matter that has existed for billions of years.

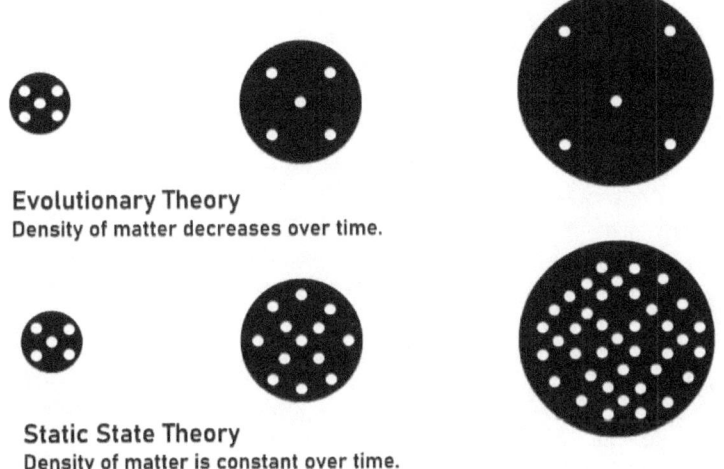

Evolutionary Theory
Density of matter decreases over time.

Static State Theory
Density of matter is constant over time.

In 1968, the Orbiting Astronomical Observatory 2 became the first successful space telescope. Since then, we have launched over 100 telescopes into space, and none of them have found traces of new matter being created on its own. Today, we know that a limited amount of matter is present in the universe. As the universe expands, galaxies drift apart, eventually reducing the overall density of the universe. Also, based on how far we can glimpse into the universe, it appears that our universe has a finite age with a beginning that can be explained. We call it the Big Bang theory.

The Big Bang

Let me ask you a simple question: How old are you? Really, how old are you? The simple answer is the number of years from the year of your birth. That is the human way of looking at it. In a more cosmic way, we are billions of years old because that is how old all the atoms in our bodies are.

Edwin Powell Hubble was an American astronomer. While observing distant galaxies in the 1920s, Hubble discovered something that changed the course of astronomy. He found that the light coming from two galaxies, located near and far away from Earth, was not the same. The light from a galaxy farther away from Earth was more shifted toward the red spectrum than a nearer one. In simple words, the wavelength of light was increasing; it was being stretched before it reached our planet. This phenomenon is known as redshift.

This discovery made him think that something must be happening between Earth and those galaxies, causing the redshift effect. So, what did he do? He observed more galaxies and, surprisingly, noticed the same phenomenon with every galaxy he pointed his telescope at. He knew that the space between Earth and those galaxies was increasing. In other words, he discovered that the universe was expanding. As a result of this expansion, by the time light reaches Earth, it is shifted toward the red end of the spectrum. Hubble published his findings in a paper in 1929. At that time, astronomy was not considered a part of physics, making him ineligible for a much-deserved Nobel Prize.

Just two years later, a Catholic priest and physicist, Georges Lemaître, coined the idea of the Big Bang. His reasoning was simple: if, at this point in time, the universe is expanding, then when we go back in time, the universe must contract. Furthermore, if you contract the universe enough, it becomes a point—a point of infinite potential.

Let us go back 13.8 billion years in time to understand the universe's origin. Everything that exists today, everything that we know, and everything that we will ever know was lying somewhere in a tiny bundle of energy. This bundle of energy was smaller than an atom and infinitely dense. We like to call it a point of singularity. The energy required to give birth to an entire universe that will have trillions of stars and billions of galaxies was contained within this point of singularity. One day, a species born from the same energy released from the singularity will write about it—because this infinite energy point was not stable at all.

When the singularity acquired enough randomness, the Big Bang explosion occurred. This explosion gave birth to both space and time. It was more powerful than any explosion we have ever witnessed or will ever see. Within a fraction of a second, the Big Bang created all the energy that would fuel billions of galaxies and give birth to trillions of stars. Things get hot when there is an explosion. For comparison, the temperature at the core of our Sun, where all the fusion takes place, is 15 million degrees Celsius. When the Big Bang explosion occurred, the immediate temperature was over 1,000 trillion degrees Celsius—hotter than anything we have ever seen. The closest we have gotten to this temperature is inside the Large Hadron Collider, and it is still below 10 trillion degrees Celsius.

One of the fascinating things about the energy released after the Big Bang is that it can be converted into different forms. Let me ask you—what is common between a pen, a smartphone, the people you love, and the very planet you walk upon? Every single thing we can see or touch is made from the different forms of energy released after the Big Bang. We are all nothing but a naturally organized heap of the same condensed form of energy. The Big Bang explosion was so powerful that our universe went from a size smaller than an atom to larger than a galaxy in a short period of time. As the universe expanded exponentially, it also cooled down. When the universe was cold enough, electrons, protons, and neutrons were formed from the same energy released in the Big Bang.

There was a time when we thought that the Earth was the only planet in the universe—and look where we are today. Today, we cannot go 13.8 billion years back in time and observe what happened. Even if we could go back in time, we would not be able to observe the singularity because time and space did not exist back then.

Foundations of the Big Bang model

There are foundations upon which every scientific theory stands. When it comes to the Big Bang model, we have multiple pieces of evidence that it happened, and here are some of them. The Big Bang theory states that lighter elements such as hydrogen and helium were formed in large quantities after the explosion. It was easy for single or double protons and neutrons to bump into each other and sometimes even stick together, forming the nucleus of an atom. Today, when we observe the universe, we find that approximately 73% of the total mass is hydrogen, and another 25% is helium. Almost all the stars in the universe, including our own Sun, are made up of hydrogen and helium.

Whenever we talk about the Big Bang theory, it is essential to talk about the cosmic microwave background radiation because it confirms the essence of this theory. Moreover, what might be one of the most significant revelations of the 20th century started with bird droppings.

Cosmic Microwave Background Radiation

Arno Penzias and Robert Wilson are two American astronomers. In the early 1960s, they set out to map radio signals from the space between galaxies. Nevertheless, no matter where they pointed their telescope, there was a constant background noise interrupting their observation. They cooled down the receiver with liquid helium to eliminate all interference, but it did not help. Then they thought it might be due to the birds nesting on the horn-shaped antenna; they spent hours cleaning the antenna and removing bird feces. Even then, they had constant static, radio-like noise from all directions.

The antenna was located close to New York City in Holmdel. So, they decided to point the telescope at New York City to ensure that it was not coming from the city either. After eliminating all the known sources, they knew they had something on their hands. Penzias and Wilson quickly began looking for a theoretical explanation for what they saw. At the same time, the physicist Robert Dicke theorized that if our universe was superhot at its beginning, it must have created a lot of radiation. That same radiation would still exist throughout the universe. He visited Penzias and Wilson at Bell Labs and confirmed their findings. The mysterious signal was the cosmic microwave background radiation left over from the Big Bang. Penzias and Wilson won the 1978 Nobel Prize in Physics for this discovery. By using modern telescopes, astronomers have mapped out the entire observable universe and created a map of the CMB radiation.

The Flaws of the Big Bang Theory

The Big Bang theory has been wildly successful so far. However, no theory is perfect, and there are things that even this model cannot explain. There are some areas that the Big

Bang model has left untouched. The Big Bang model states that before the Big Bang, the universe was infinitely dense, stuffed into a point we call a singularity. So, where did the singularity come from? Did it come from an event that we do not know yet, or has it always existed?

We have come to the conclusion of a singularity from the fact that our universe is expanding today. But there is an even bigger question: Was there a singularity? How can something—the singularity—come out of nothing? The closer we cruise to the Big Bang, the laws that govern our universe seem to break down. We are not sure if a singularity is the only reasonable explanation for our universe's birth.

Let us suppose that our universe did come from a singularity that came out of nowhere. Let us assume that the Big Bang model is the only true universe model. Why did it bang? If the singularity had existed for eternity, what forces acted upon it that caused it to become unstable and explode? Was that force internal? Did the singularity become unstable from the inside, or were there some external forces? We can only speculate because we have no way of knowing that today. If we can figure out where the singularity came from and why it exploded, everything else will slowly fall into place.

Age of the Universe

Today, scientists can confidently say that the universe is 13.8 billion years old, with an uncertainty of less than 1%. But how do we know the precise age of the universe? Well, we have two methods to determine it.

First of all, we have Hubble's Law. Hubble's Law states that the rate at which a particular galaxy moves away from us is directly proportional to its distance. This means that galaxies farther away from us are moving faster. If this expansion continues, there will be a time when we will no longer see those galaxies because the expansion rate will be faster than the speed of light, and their light will never reach us. By calculating the expansion rate based on their distance, Hubble estimated how long ago galaxies started moving apart. Initially, Hubble underestimated the distance of galaxies and figured that the universe was about 2 billion years old. Today, we know his calculations were significantly off.

The second method involves measuring the age of the oldest stars in the universe. The first stars in the universe were formed from gas clouds about 150 to 200 million years after the Big Bang. Suppose we can point our telescopes toward the oldest stars and measure their age. In that case, we can also get a rough estimate of the universe's age. This method is a bit more complicated, as it is not easy to find a star that old. Most stars die out within their first 10 billion years, depending on their mass. A star like our Sun can live up to 10 billion years, whereas a star 20 times its size can only live about 10 million years. Even if we can find such a star, it is not easy to precisely measure its age. For example, the oldest star we have found in the universe is the Methuselah Star, or HD 140283. Scientists have estimated its age to be 13.7 billion years, with an uncertainty of 700 million years. If the Methuselah Star is younger than the universe, that is fine. But the uncertainty says that it could be as old as 14.4 billion years, taking us back to the drawing board.

There was a time when the Steady-State theory seemed to have all the answers to our questions concerning the origin of the universe. As we looked deeper, our understanding changed. Naturally, the Big Bang theory replaced the Steady-State theory. For decades, the Big Bang theory has been on the front line of our exploration. How long do you think it will be before another theory—or a better version of this one—comes to light?

Even today, we have theories that challenge or at least add something to the Big Bang model, such as the Eternal Inflation theory and the Oscillating Universe theory.

In a way, the Eternal Inflation theory is an extension of the Big Bang theory. It suggests that our universe went through a rapid expansion for a brief period called inflation after the Big Bang. This inflation did not stop even after a few billion years—and it never will. The inflation will go on for an infinite period, or as long as the universe exists. Today, we can observe this inflation in the form of the expanding universe.

The Oscillating Universe theory is a bit more complicated. Let us suppose you have a spring in your hand. When you stretch the spring and release it, it oscillates. According to this theory, our universe goes through an endless series of stretches and contractions, just like the spring. The universe begins with the Big Bang and expands. Once it has reached the maximum possible expansion, the force of gravity will take over, and the universe will start to contract. Eventually, it gets to the point of singularity and—Bang—a new universe forms from the older one. The cycle continues.

The Steady-State theory is a theory of the mid-1900s that has been ruled out by our current understanding of the universe. Even though the Eternal Inflation theory and Oscillating Universe theory are still popular among scientists, neither of these theories is as successful as the Big Bang theory.

Universal Forces

Whenever scientists discovered a new force, it changed history. Sir Isaac Newton first came up with the theory of gravitation in the 1660s. Once this force of the universe came to light, it revolutionized science. People started wondering: if Earth's gravity keeps us on this planet, then what impact might it have on other objects in space? The discovery of electromagnetism in the 1820s brought light to this world and revolutionized the modern world we live in. Today, we know there are two more fundamental forces of the universe, namely the strong nuclear force and the weak nuclear force.

If you see an atom, you will find order there. There is a tightly held nucleus at the center and electrons in a particular orbit. If you look at the solar system, there is order there. There are planets, asteroids, and comets circling in a specific orbit. There are systems in our universe that seem chaotic and unpredictable. But if you look closely and give them enough time**,** you will see that there is order there. We have order because fundamental forces govern whatever happens in the universe. Today, we know these fundamental forces in the form of gravitation, electromagnetism, strong nuclear force, and weak nuclear force. Let us look at them closely, one by one.

Gravitation

Gravity is the weakest force in the universe, yet it has an infinite range. Suppose you and your friend are standing at opposite ends of the universe. Both of you would still be attracting each other with gravity. However, its strength would be almost negligible. The nature of gravity is that it is always attractive.

Sir Isaac Newton and Albert Einstein described this force in their own ways. Sir Isaac Newton said that objects experience gravity because they have mass. The more mass an object has, the greater its gravity will be. Later, Albert Einstein made significant contributions to this matter. He explained the entire universe as a sheet of fabric distorted by the mass of stars and planets. This distortion causes objects to fall toward larger objects. In simple words, you can stand on planet Earth not because Earth is pulling you toward itself**,** but instead because space is pushing you toward Earth. The same force also binds solar systems together.

Electromagnetism

Electromagnetism is derived from two separate words: electricity and magnetism. Earlier, it was believed that electricity and magnetism were two independent forces. However, when we study charged particles, we find that they are actually the same thing. If you have two oppositely charged particles—positive and negative—there is an attractive force between them. If the electric charges are the same, the force is repulsive. At the same time, if you have two magnets, the north and south poles would attract each other, and like poles would repel.

The relationship between electric and magnetic forces is simple: if we move a magnet, we can generate a charge. Furthermore, if we move a charge, we can generate a magnetic field.

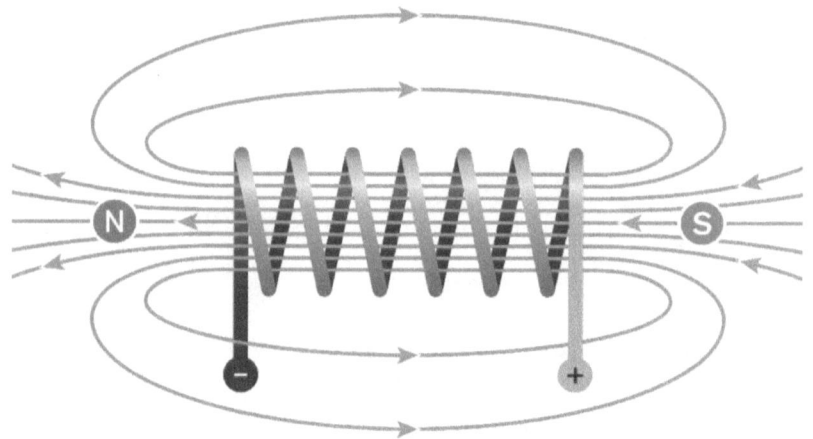

The electromagnetic force has illuminated our entire world since its discovery. This discovery gave birth to electronic appliances. All the electronic gadgets in our homes—such as laptops, personal computers, lights, and TVs—are based on this force. The computer revolution we have seen in the last 50 years would be incomplete without it.

Strong Nuclear Force

Have you ever wondered why the gold jewelry you wear does not just fall apart? Why do the atoms of gold—more specifically, the nuclei of gold with 79 protons—stay intact? The same charges are repulsive in nature, and gold atoms have 79 identically charged protons in their nucleus. In theory, there must be chaos there. It turns out there is a force that binds the hearts of all the atoms in our universe together. This force is known as the strong nuclear force, or the strong interaction.

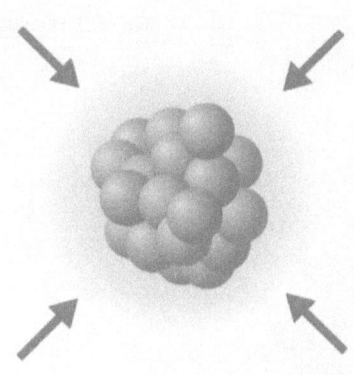

This force not only binds the atomic nuclei, the protons, and neutrons together, but also binds the quarks—the fundamental particles that make up protons and neutrons. It is impossible to detect this force directly, as it works on a very small scale. To detect the strong nuclear force, we need a particle smasher—a machine that smashes particles together and breaks them apart into their constituents. Once a particle is broken, we can study the trail it leaves behind and understand what it is made of.

Weak Nuclear Force

There are three quarks in both protons and neutrons. Protons consist of two up quarks and one down quark, whereas a neutron has one up and two down quarks. The strong nuclear force binds these quarks together, whereas the weak nuclear force allows them to change type. For example, changing one up quark into a down quark turns a proton into a neutron, which results in the decay of an atom.

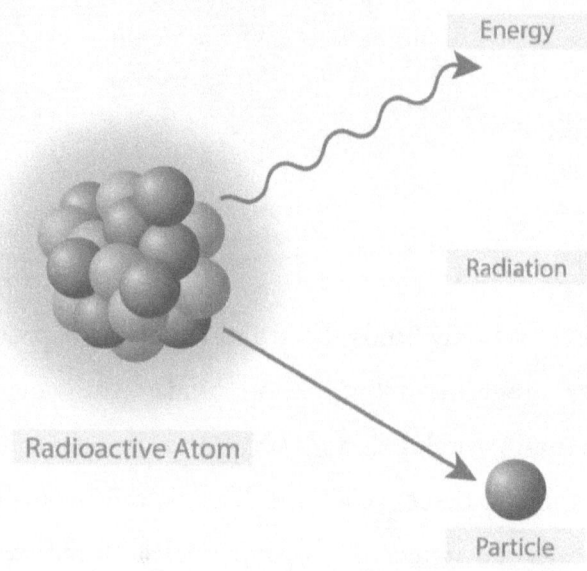

The weak nuclear force makes nuclear fusion at the core of our Sun possible. It also plays an essential role in nuclear fission, used in nuclear power plants, where radioactive elements such as uranium are used to generate electricity.

The Story of Light

Fear is necessary for the survival of a human being. Imagine standing on the edge of a mountain cliff and not being afraid of falling down and injuring yourself. Would you not want to try seeing what happens when you go down? After all, it is just an unexplored territory. When a baby is born, they have two fears—the fear of falling and the fear of loud noises. The fear of falling prevents a baby from getting severely injured. The fear of loud noises helps them alert their parents by crying when they hear something unusual. There are other fears that we develop as we grow up, such as the fear of the dark.

Humans fear what they cannot see. From an evolutionary point of view, it is natural for humans to be afraid of the dark. The best way to eliminate any fear is to understand what you are afraid of. In the language of science, darkness is nothing but the absence of photons. There is absolutely nothing to be afraid of in the absence of photons. A photon is a particle of light that carries electromagnetic radiation. A tiny portion of this radiation is called visible light, which allows us to see the world.

In his theories, Sir Isaac Newton predicted that the speed of light could be changed. It is variable depending on the frame of reference. For example, if you have a torch in your hand and you are running in the direction of an object with a velocity 'v,' the full speed of light would be equal to the speed of light (c) plus your running speed—c + v—and vice versa. Newton believed that the entire universe was filled with a hypothetical medium called ether—a massless, invisible, and infinitely low-density medium. Ether provided a rest frame for the propagation of light. The high value of elasticity and very low density enabled light to propagate through this medium without losing its intensity. Not only light, but this ideal medium also enabled electromagnetic waves to travel through space.

To check the existence of ether, one of the most significant and long-lasting experiments in human history was carried out: the Michelson–Morley experiment. The basic idea of this experiment was to calculate the relative motion of light with respect to Earth. If there was a medium called ether in space with low density, there must be a relative motion. In the end, no relative motion between Earth and ether was detected. This experiment won Michelson the Nobel Prize in Physics in 1907.

Albert Einstein strongly believed that the speed of light is invariant and independent of the frame of observers. c + v and c − v are the wrong assumptions, and it is always c no matter what frame you are in. An object cannot outpace the speed of light, even if you continue to apply force. If an object reaches the speed of light, its energy would become infinite, making it impossible to push further.

We have a star—the Sun—as our primary source of energy and light. The presence of light has transformed our dark universe into an observable one. In the absence of light, we would not be able to do most of our space-related experiments. Our telescopes rely on the presence of light, and without it, they would become useless.

The speed of light

The speed of light has been a major discussion among scientists for centuries. In the 17th and 18th centuries, scientists were divided into two categories. One group believed that the speed of light is too fast—or infinite—so it can never be measured. At the same time, the other group believed that light travels at a finite speed; therefore, it can be measured using scientific instruments.

In 1676, astronomer Olaus Roemer conducted one of the earliest experiments to calculate the speed of light. Roemer was working at the Paris Observatory. His primary role was to observe the orbit of Io, one of Jupiter's moons, and accurately time the eclipses. He observed Io for many years and found something he was not even looking for.

Roemer noticed that, in its orbit, as Earth moves toward Jupiter, the eclipse would start early. Likewise, it comes at a later time when Earth is moving away from Jupiter. Roemer calculated that the eclipse occurs 11 minutes earlier than its predicted time when Earth is closest to Jupiter, and 11 minutes late when it is at the farthest point. He realized that this time difference must be due to the finite speed of light. When Earth is near, light reaches us faster; and later, when it is farther away. He estimated that, overall, light takes about 22 minutes to cross the entire orbit of Earth around the Sun. By dividing the diameter of Earth's orbit by the time taken by light, he estimated that the speed of light must be around 220,000 kilometers per second. Even though he was way off, he dismantled the myth that light has an infinite or non-measurable speed.

In 1726, James Bradley measured the speed of light by using the method of stellar aberration. To understand stellar aberration, let us use falling rain as an example. Suppose you are standing still in the rain with no wind. The droplets will fall vertically and hit you directly on the head. However, if you start running, the rain will hit you on the front at a specific angle, depending on your running speed. As we know, our planet revolves around the Sun, so the position of distant stars must change slightly due to its yearly motion. Knowing the speed of Earth around the Sun, Bradley measured this angle for starlight and calculated the speed of light to be about 301,000 km/s.

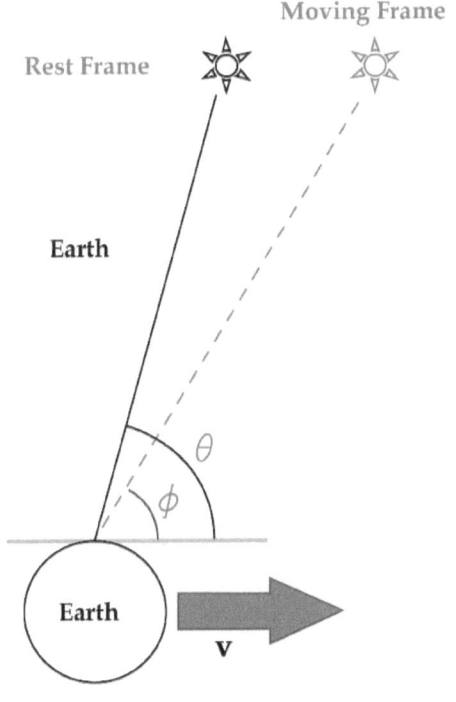

In 1849, Armand Fizeau measured the speed of light to be 315,000 km/s using the toothed wheel method. Fizeau used a beam splitter and focused the beam of light onto a plane mirror, where a spinning toothed wheel was located. Light passing through the toothed wheel was projected to a mirror located 5 miles away. The reflected beam of light was then returned to the point of origin. Fizeau kept increasing the speed of the rotating wheel until the tooth of the wheel entirely blocked the returning light from 5 miles away. Knowing the wheel's speed and the distance light had traveled back and forth, Fizeau measured the speed of light to be 315,000 km/s.

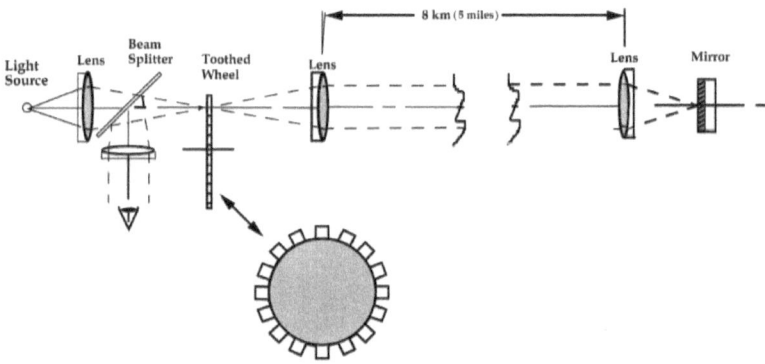

In 1862, Léon Foucault measured the speed of light to be 298,000 ± 500 km/s using the rotating mirror method. Léon Foucault made one of the most accurate measurements of the speed of light in his time. His experiment consisted of two mirrors—one fixed mirror, while the other was rotating. The basic technique here was to send a sharp beam of light on a path to bounce between a rotating mirror, a fixed mirror, and then back to the rotating mirror, for a total distance of 2D.

As light traveled the 2D distance and came back, the rotating mirror would have turned very slightly in angle. This slight rotation deflected the beam of light through a small angle θ from its original path, producing a measurable effect. Using this idea, Foucault calculated the speed of light to be 298,000 ± 500 km/s, which is very close to what we know today.

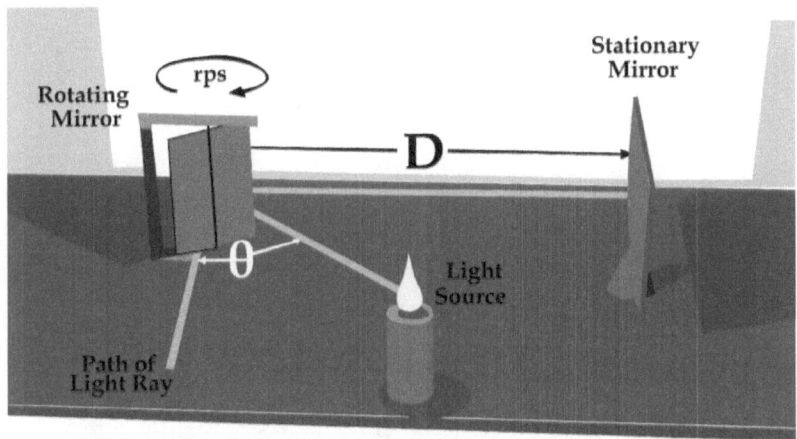

In 1958, K.D. Froome measured the speed of light to be 299,792.5 ± 0.1 km/s using the radio interferometer method. Interferometry can also be used to measure the wavelength of electromagnetic radiation, which can be very useful when determining the speed of light. At that time, laser technology had not been invented, so Froome used other coherent radio sources to measure the speed of light.

In 1973, Evenson et al. measured it to be 299,792.4574 ± 0.001 km/s using lasers. It was the most accurate value because lasers are highly monochromatic and unidirectional.

Finally, in 1983, the speed of light was accepted by the scientific community worldwide to be 299,792.458 km/s. Roughly, we use it as 3×10^8 m/s in our mathematical calculations. The real credit for calculating the speed of light goes to Olaus Roemer, who was able to estimate its rough value by observing the motion of planets—even when the distances between planets were not accurately known.

What is a light-year?

A light-year is often misunderstood as a unit of time. Instead, it is a unit of distance. One light-year is the distance traveled by a beam of light in one year—equal to 5.8 trillion miles or 9.4 trillion kilometers. The diameter of our Milky Way galaxy is about 100,000 light-years. If light starts traveling from one end of the Milky Way, it will take 100,000 years to reach the other end.

Time Travel

Is it possible to travel in time? The straightforward answer is: we are all time travelers. Time is one of the fundamental properties of the universe. Even as you are reading these lines, you are traveling in time. Time is taking us into the next moment after every present moment. The universe itself governs this "time machine" and has its own set of laws. That is why we consider time to be the fourth dimension of the universe. After every present moment, we are heading into the future, and the past becomes a memory.

If we are all time travelers, what is time travel?

Time travel is about going into the future or past moments at a faster or slower rate. Alternatively, simply switching between different points in time is time travel. Traveling into the future has already been proven experimentally, as scientists have observed particles that appear to experience time differently. However, time travel into the past is a bit more complex and requires further understanding.

Into The Future

Practically, it is impossible for humans to travel in time right now. Theoretically, however, there are multiple ways to travel into the future. Some of the possible explanations for forward time travel include:

High Velocity: Suppose there are twin brothers, Sam and Leo, aged 24. One day, Leo decides to enter a very high-speed spaceship and travel to Proxima Centauri. It is the nearest star to Earth after the Sun, located 4.2 light-years away. Leo's spaceship is very advanced and can reach a speed close to the speed of light. Leo enters this spaceship and programs it to travel toward Proxima Centauri at 99.9% the speed of light. Once he reaches the star, Leo decides to return to Earth.

During this whole time, Sam stays back on Earth. From Sam's frame of reference, over 8 years have passed since Leo took off, whereas for Leo, only a few months have passed. Sam is now over 32 years old, whereas Leo is only a few months older. This phenomenon is called time dilation. Time dilation states that time slows down as you approach the speed of light, compared to someone in a stationary frame of reference.

In order to make such time travel possible, we need high velocity. Today, we do not have the technology—or the spaceship—that can travel close to such speeds. Our fastest spaceships can reach only a fraction of the speed of light. We have made significant leaps in technology in the last 50 years, and we can hope to make it possible in the coming 1,000 years. Until then, we can explore other means of time travel, such as using a black hole.

Using Black Holes: The idea of using black holes to travel forward in time comes from our understanding that time slows down when you are near an object with high gravitational potential. Gravity has the ability to bend the fabric of space-time. The stronger the gravity, the more it bends the fabric of space-time, resulting in a slower passage of time. What in the universe has a higher gravitational potential than black holes?

If we take a very advanced spaceship and orbit around a black hole, time will run slower for us than for a person on Earth. Do not get confused when I say that time will run slower. Time running slower does not mean things will happen in slow motion for us in the spaceship. Our feeling of time will remain unchanged. Instead, what will happen is that after we revolve around a massive black hole for one month and return to Earth, several months would have passed on Earth. Our feeling of time and the way we experience it will remain unchanged. We might not even notice that we have traveled in time. This is the second form of time dilation, also known as gravitational time dilation.

There are two main problems with using a Black Hole as our time machine:

The first problem is that black holes are extremely powerful and have immense gravitational potential. Most black holes can consume an entire star system, so our spaceship must be fast and powerful enough to escape. Just being fast enough will not work near a black hole. Our spaceship must also be strong enough to withstand its gravity.

The second problem is the human body in that spaceship. We have all grown up on Earth, experiencing a specific gravitational field. If you try to stand on the surface of a neutron star, you will get crushed almost immediately—to the level of an atom. We have already seen examples of the human body acting differently in different gravitational fields. The astronauts who go to the ISS (International Space Station) and spend a few months in zero gravity get stretched. Their height increases by a few centimeters. Living here on Earth, we are continuously pushed toward this planet. How would that impact the astronauts when our spaceship gets very close to a black hole? Would they be able to stay alive while being in such an intense gravitational field?

Both of the above scenarios seem impossible to overcome in the short term. Our nearest black hole is A0620–00 (V616

Mon), which is located roughly 2,800 light-years away—making it incredibly difficult to reach.

Time travel is not impossible, but that does not mean it is possible either. Time dilation has been scientifically proven in measurements of atomic clocks. One of the most well-known real-life examples of time dilation is the μ-meson. μ-mesons are unstable cosmic rays formed 10 km above Earth. Their lifetime is about 2×10^{-6} seconds, and their velocity is about 0.998c. In theory, μ-mesons should travel a distance of about 600 meters before disintegrating. But here comes relativity: their lifetime increases slightly because of time dilation, and they can be found all over the Earth.

For some physicists, time travel is a human construct, and what we are really doing is manipulating a property of space-time—especially when we are moving forward in time. Time may flow in a definite direction, and we may travel forward in time, but traveling back seems impossible. It is like a game where the door behind you closes once you've taken a step forward.

Into The Past

Let us begin with a small story. Suppose Max is a time traveler. He travels back in time to the 1950s, when his grandfather was a handsome young man. He steps into his grandpa's home and finds that it is the day his grandfather is going to meet his grandmother for the very first time. Max meets his grandfather and somehow stops him from meeting his grandmother. In doing so, his grandfather will not be able to meet his grandmother. This means Max's father will never be born in the future, and thus, there will be no Max. In this paradox, Max prevented his father's birth and, consequently, his own. However, this cannot be true in the real world, because Max is already present in the future. How can two opposite events occur at the same time? How can Max be alive when his parents were never born? Does that mean we cannot travel back in time? This problem is known as the grandfather paradox.

One theory suggests that even if we could travel back in time, we would not be able to influence the future. From the point in time you traveled back from, the future will go on its path without any influence from time travel. Physicists explain time and traveling back in time using the example of a river. Time is like a river flowing in one direction, and we are sitting in a boat on that river. This way, all of us are experiencing the same time. Now suppose someone jumps out of the boat and tries to swim backward. They will not be able to do so. However, let's assume they do swim backward. In that case, they will create a separate timeline of their own. In the new timeline, the consequences of their actions will occur, but the future they came from will remain untouched.

So, if Max travels back in time, he will create a new timeline. In this timeline, his father will never be born, and things will go as they should. Some scientists suggest that it is impossible to create a separate timeline. If we somehow travel back in time, we will not be able to influence it. This means that Max will not be able to stop his grandfather from meeting his grandmother. This idea raises questions about the nature of the grandfather paradox. Also, why should we travel back in time if we cannot interact with it or have any influence?

There is one more idea that revolves around the grandfather paradox. Whenever Max goes back in time and modifies his past, he creates a new universe. This way, he stops the meeting of his grandparents in a universe where he is unknown. Backward time travel is not just difficult but also very confusing.

One of the best arguments against the possibility of backward time travel is that we do not have any humans from the future. If it were humanly possible to travel back in time, why haven't future humans come and shared all their technological secrets to make things easier for us? However, the counterargument is that maybe we are the future humans. Maybe we are in the front seat of the cosmological timescale. We are the ones who will go back in time—since we have had a past.

Multiverse

There was a time when Earth was considered the center of the universe. This is not surprising because, at that time, when we looked up at the night sky, all the stars appeared to be moving around Earth. So, it was natural to assume this 600 years ago, especially with the limited knowledge presented in religious texts.

It was also a common belief that we are the only living creatures in the universe and that there is zero possibility of life on any other planet. To some extent, this belief still persists, because we have not found alien life yet. But some things have changed. We have found that we are not the center of the universe. There are billions of planets orbiting the habitable zones of their respective stars. The possibility of finding alien life has only gone up in the last few decades. We have become more open-minded to new ideas over time—especially the ones we can observe.

As per the discoveries made in the 19th century, our universe was once thought to be the only one. Everything that we know or will ever know was believed to be confined within the boundaries of a single universe. But today, we have theories describing the possibility of multiple universes—i.e., a multiverse. Multiverse theories suggest that our universe is not alone. There is a series of multiple universes, and our universe is just one among a finite—or perhaps infinite—number of them.

One of the biggest pieces of evidence for the multiverse is our own existence. We know at least one universe exists—our universe. So why can't there be others? The presence of our universe doesn't answer the question, but it opens the possibility. We have theories that support the idea of multiple universes, such as string theory. It suggests that there could be other universes and that we might one day move between them. When two universes collide, they may fuse into one, forming a much larger universe. When a single universe splits, it creates two independent universes with their own laws and properties. The multiverse seems strange to imagine because this is the only universe we have ever known.

It is predicted that our universe is part of a much bigger picture. We are nothing but a drop in the ocean. We don't know the easy answer to this question because we have no way of finding out for ourselves. Some scientists believe that multiple universes originate from an enormous ocean of energy foam. The birth and death of a universe in this energy foam are like bubbles appearing and disappearing in a bathtub. Some of these bubbles break almost instantly, while others last for a while. The idea of our universe in an energy foam is fascinating. However, it takes us back to the first piece of the same puzzle: if our present universe is part of a much larger ocean, then where is that ocean? Inside an even bigger entity? Who knows.

Some physicists believe that the multiverse is just a vague term that has nothing to do with reality. If our universe is just one of many, where did the multiple universes come from?

Now you might ask: if there are multiple universes, how can we leave this universe and enter a new one? Well, this is where the idea of wormholes comes into play.

Wormholes

In the 1930s, Albert Einstein and physicist Nathan Rosen came together and proposed the idea of wormholes. Wormholes are also known as Einstein–Rosen bridges. They used Einstein's theory of relativity to conclude that "bridges" can be created in space-time. These bridges can connect long distances, effectively reducing travel time and distance.

To understand how wormholes work, we can use a simple concept. Take a sheet of paper in your hand and mark two points—A and B—at opposite edges of the page. If you want to travel from point A to point B, you can take as many routes as you want, but the shortest one would be a straight line. Suppose the distance between points A and B is around 15 cm. Starting from point A, you travel at 1 cm per second. It would take you 15 seconds to reach point B. Now take the same piece of paper and fold it so that the two points touch each other. The distance between A and B becomes almost zero, and you can travel from A to B in virtually no time..

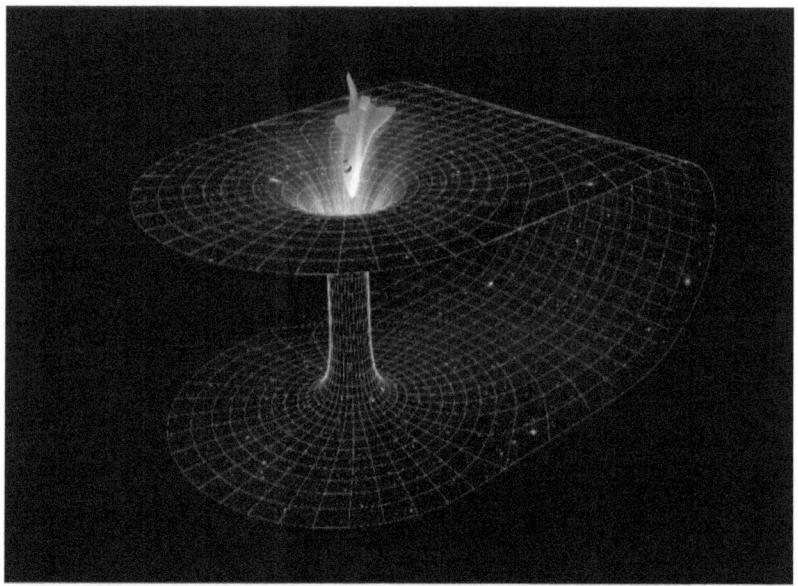

Wormholes can also be thought of as pathways or tunnels within one or more universes. As we know, matter and energy can warp the fabric of space-time. But in a particular configuration, this warping could act as a tunnel, allowing the passage of matter or information. Creating a wormhole would require an immense amount of energy and tools that we do not possess today. However, some physicists suggest that wormholes may exist naturally in the universe. If we look deep enough, with the right tools in the right places, we may even find one. Instead of creating a new wormhole, we could potentially use an already existing one.

With our current theoretical understanding of wormholes, it is not clear if we can pass through them without tearing them apart. It is still debated how a wormhole would react when a piece of matter passes through it. Alpha Centauri is located around 4.3 light-years away. Practically, if we started traveling from Earth in a spaceship, we would never reach it within a human lifetime. It would take over 70,000 years of continuous travel just to get there. Wormholes could come in handy for such journeys. We could create a wormhole whose one end is in our solar system and the other opens near Alpha Centauri.

What is the difference between a Black Hole and a Wormhole?

Black holes are more like cosmic vacuum cleaners. They take in anything that comes their way. Their gravity is so strong that nothing can escape. Black holes are formed by massive amounts of matter collapsing under their own gravity, resulting in a singularity.

Wormholes, on the other hand, are theoretical shortcuts between two distinct parts of the universe. Instead of ending in a singularity, a wormhole opens up at another location in space-time.

Today, we have observational evidence of the existence of black holes, whereas wormholes remain purely theoretical.

PART-II

After the Big Bang

There was a time when scientists believed that our universe consisted only of matter—matter that you can touch, see, and feel. It was a common belief that there was nothing in between the galaxies. As a result, matter was the only point of discussion among scientists before the 20th century. The study of matter helped us immensely. We discovered a large number of subatomic particles, which gave birth to quantum mechanics.

But as we looked deeper, we realized that there are forces in the universe that we haven't fully understood yet—forces acting on galaxies and keeping them together. Without them, most galaxies would fall apart. We named this mysterious force dark matter.

Our quest to understand the expanding universe has taken us beyond the Milky Way galaxy and toward the edges of the cosmos. We realized that there is another force driving the universe's expansion, and we call it dark energy.

Matter

To understand matter and its nature, let us do a simple experiment. Let's reverse the clock and go back to the Big Bang explosion. As we move backward in time, we become apes. We see the dinosaurs roaming around planet Earth. As we keep going, we find ourselves in the ocean as aquatic life 500 million years ago. 4.5 billion years ago, we witness the formation of our planet, and around 13.6 billion years ago, the formation of our galaxy. We're getting close. As we approach the Big Bang explosion, we find ourselves in the form of early atoms falling apart. Soon, our atoms completely disintegrate into pure energy. Eventually, we arrive at a tiny singularity— the same singularity where it all began. At that point, there will be no space for us, and time will stop. We will not be able to go any further back.

In the first moments of the Big Bang explosion, the universe was too hot for anything to form. But as the universe expanded exponentially, it also cooled down. At a lower temperature, the conditions became right for the formation of the building blocks of matter—quarks and electrons. A few millionths of a second later, once the universe reached a specific temperature, quarks came together and arranged themselves, forming the first protons and neutrons.

A few millionths of a second after the Big Bang, we had a universe full of subatomic particles: protons, neutrons, and electrons. Scientists describe it as a superheated soup with all the ingredients to form matter. But there was one problem—it was still too hot. Even if protons and neutrons tried to interact, the heat would rip them apart.

About 3 minutes after the Big Bang explosion, the universe was cold enough for these protons and neutrons to form the first nuclei. Now, our universe was full of atomic nuclei roaming around. Only one last ingredient was missing to create the first atoms: the electrons. Even after hours, days, and years, the nuclei couldn't catch the electrons. The heat of the universe always prevented them from doing so.

It took another 380,000 years of expansion and cooling for these nuclei to finally capture electrons and form the first atoms—the building blocks of life and everything around us. The very first atoms that were created were hydrogen atoms. Hydrogen requires only one proton and one electron to form a stable structure. There are other hydrogen isotopes, such as deuterium and tritium, with one and two extra neutrons, respectively. But after deuterium, hydrogen becomes unstable and doesn't last long. Due to the simplicity of its formation, hydrogen was created in abundance in the early universe.

Another element that formed, though not as abundantly, was helium. Two electrons are bound by two protons with either one or two neutrons. This composition can still be observed in the universe today. Almost all the stars we see in the night sky are mostly made up of hydrogen and helium. The early universe was like a big cloud of fog. But with the formation of atoms, it became more and more transparent.

So, how did we go from a universe full of atoms to one filled with stars and galaxies? The simple answer: gravity.

After the formation of the first atoms, over the next millions of years, gravity did its magical work. It pulled together the unevenly scattered atoms and formed large gas clouds across the universe. As these gas clouds grew, their collective gravity attracted more matter. Smaller gas clouds merged with larger ones, resulting in massive gas clouds—and the process continued.

From a distance, everything looked calm. But if we moved to the center of these gas clouds, something else was happening. In the womb of these clouds, the very first stars were being born—thanks to the massive gravitational forces at play.

Hydrogen and helium atoms were being forced to fuse. At temperatures of hundreds of millions of degrees, hydrogen was fusing into helium. Helium then fused into heavier elements like lithium. In this process of combining smaller atoms into bigger ones, energy was released. That energy became the glow of our newly born stars.

The early stars were massive—there was little competition and too much matter. The universe twinkled with their light. However, the larger the star, the faster it fused matter at its core, reducing its overall lifespan. At the end of their life cycles, most of those stars exploded as supernovae. They gave us heavy elements such as carbon, oxygen, iron, gold, and more. The carbon in your body and the gold in your jewelry were formed in the cores of dying stars billions of years ago.

Some of those stars even collapsed into black holes, becoming the guiding centers for the formation of the first galaxies.

Anti-matter

No story is complete without a good rivalry. In our story, matter and antimatter had the biggest rivalry in the history of the universe. When they meet, matter and antimatter completely annihilate each other, leaving behind pure energy. In the moments after the Big Bang, our universe was filled with protons, neutrons, electrons, and their corresponding antiparticles. For every proton, there was an antiproton; for every neutron, an antineutron. For every electron, there was an anti-electron—also called a positron.

What is antimatter, you might ask? It is the opposite of ordinary matter. Antimatter is composed of antiparticles that have the same mass as ordinary particles but the opposite charge. In the early universe, just like normal matter, antiparticles came together to form antimatter. An antineutron and an antiproton combined to create an antihydrogen atom, with a positron in its orbit.

The Big Bang explosion is believed to have created equal amounts of matter and antimatter to preserve the universe's symmetry. But under normal circumstances, these two should have annihilated each other completely—leaving behind a universe filled only with energy. Yet, wherever we look, everything appears to be made of matter. There are no observable traces of antimatter left.

So, there must be unknown processes—ones we don't yet understand—that tipped the balance and resulted in a universe dominated by matter.

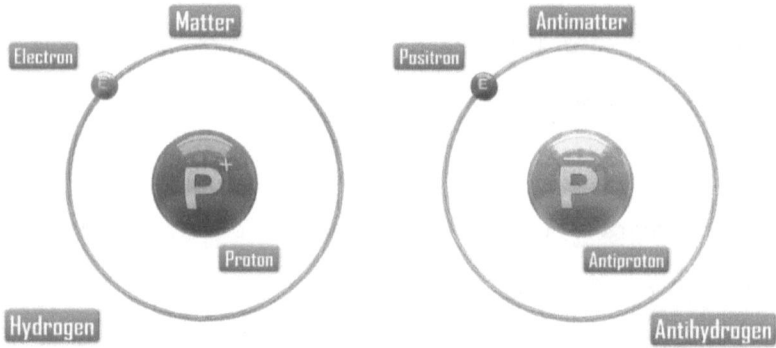

Some scientists believe that our universe isn't perfectly symmetric. Under exceptional circumstances, it could have formed different amounts of matter and antimatter. But even then, the net difference would be too small to explain all the matter we observe in the universe.

We know from experience that most matter particles are highly stable—they can exist for billions of years. The expiry date on a water bottle is not for the water but for the plastic bottle itself. After a certain period, plastic releases chemicals that can change the smell and taste of the water. The water inside, however, is billions of years old and can stay like that for billions more.

There are a few exceptions when it comes to matter that isn't very stable—such as uranium, plutonium, thorium, and radium. These elements decay slowly over time.

One theory suggests that antimatter particles are highly unstable and decay quickly. Their decay process begins the moment they are created. So, even though matter and antimatter were formed in equal amounts after the Big Bang, the rapid decay of antimatter caused it to disappear—allowing ordinary matter to dominate the universe.

In the beginning, if antimatter had won the race, our universe would likely be just as it is today—but made up of opposite particles. Our stars would be made up of antihydrogen atoms, creating antihelium at their cores. There would be planets composed of antimatter. And just like us today, intelligent beings in that universe might be curious to know what ordinary matter is and how it works.

Suppose the multiverse theory is correct and there are infinite universes. In that case, there may be some universes where antimatter won the race.

String Theory

A famous saying goes: Don't trust atoms; they make up everything. Well, people who believe in string theory might want to correct that: Don't trust strings; they make up atoms that make up everything.

In the 1950s, when scientists started smashing atoms, no one had imagined that there could be hundreds of hidden particles inside an atom. These different particles all have unique roles to play. String theory proposes that there is one more fundamental piece of matter yet to be discovered. Without this particle, the Standard Model of particle physics remains incomplete.

So, what is String Theory, and what is the basic idea?

String theory suggests that matter consists of tiny, vibrating strings of energy. Suppose you have a piece of matter in your hand—say, a potato. If you slice that potato and put it under a microscope, you will find that it is mostly made up of water, starch, and protein. Now take one molecule, like a starch molecule, and observe it under a much more powerful microscope. You would see the atoms that make up the starch molecule. Inside these atoms are protons and neutrons. These protons and neutrons, in turn, are made up of up and down quarks.

We know this much. But what are the quarks or electrons made of? What would you find if you could zoom inside a quark or an electron? This is where string theory comes into play.

String theory proposes that if you observe a quark closely enough, you will find that it is made of tiny, vibrating, one-dimensional strings of energy. These strings are the fundamental ingredients that make up everything in the universe.

You might ask: if we've already discovered so many subatomic particles, how does string theory explain their presence? According to string theory, each type of vibration of these strings corresponds to a different particle. One kind of vibration creates a quark, another produces a Higgs boson, another a muon, and so on.

All the different types of particles in the universe are nothing but different vibrations of the same string—including dark matter.

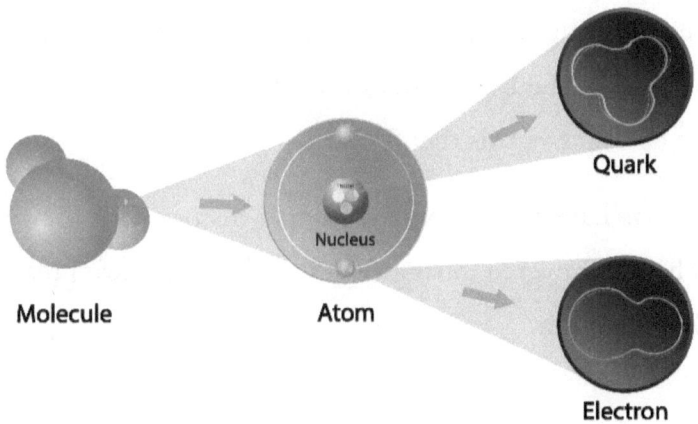

Molecule   Atom   Quark   Nucleus   Electron

String theory suggests that dark matter is also made up of the same strings as ordinary matter. However, their vibrations occur at different frequencies. As a result, we cannot see dark matter or even interact with it directly.

But how small are these strings? The strings proposed in string theory are on the order of the Planck length, i.e., 10^{-35} meters. Searching for a string inside matter would be like searching for a tiny needle in the Atlantic Ocean.

According to the Standard Model of physics, subatomic particles are considered the fundamental building blocks of matter. But string theory challenges this model—and our current understanding of atomic and subatomic particles.

String theory sounds fascinating. However, its predictions have not yet been experimentally confirmed. We currently lack the technology to probe deep enough into matter to observe these vibrating strings. Still, there is hope that particle smashers like the Large Hadron Collider (LHC) may one day prove or disprove this theory.

Physicists have revolutionized human history with the forces they have discovered. String theory could be the next great leap in that direction.

Dark Matter

Our universe is made up of three types of substances: matter, dark matter, and dark energy. It is this matter that makes us and everything we can see or feel. But there is a bigger and more dominant side of matter—dark matter. When it comes to matter, about 85 percent of the total matter in the universe is dark matter, and only 15 percent of it is ordinary matter.

So, what is dark matter, and how do we know it exists?

Born in June 1730, Charles Messier was a bright French astronomer. Messier was fascinated by comets. Every night, he would go out and observe distant comets. One night, Messier saw something that changed cosmology forever. He observed some fuzzy objects in the sky. Upon further investigation, he became sure that they were not comets.

Messier was worried that other comet hunters might get confused by these objects. So, he made a list of objects in the sky that were not comets. He talked about star clusters and spiral nebulae visible in the night sky in his list. Everyone understood what the star clusters were, as the name states. But the fuzzy "spiral nebulae" kept astronomers in the dark for the next two centuries. In the 1920s, Edwin Hubble put this matter to rest by confirming that these fuzzy "spiral nebulae" were galaxies. And there are many more of them. With this confirmation, galaxies became a point of discussion among astronomers. More and more astronomers started studying them.

During the 1930s, when scientists studied the motion of distant galaxies, their calculations did not fit with their rotation. Based on the amount of matter present in those galaxies and the speed at which they were rotating, they should fall apart. Astronomers calculated that the amount of matter in these galaxies did not have enough gravitational pull to hold them at their rotational speed. This discovery made them wonder, hmm, so what else could be holding these galaxies together? Scientists made various calculations based on the visible mass of those galaxies. Every analysis indicated one thing: there is an invisible mass whose gravitational force is holding the galaxies together. Something is there that scientists have missed for so long.

The answer came from a missing piece of the universe: dark matter—dark because we can't see it, and matter because it interacts with ordinary matter in the form of gravity. It is the gravitational force of dark matter that holds most of the galaxies in our universe together. If dark matter were to disappear from our universe today, our own Milky Way Galaxy would fall apart. Most stars would lose their orbits and scatter in the universe.

Dark matter does not emit any electromagnetic radiation, so it's impossible to see it directly. So, how do we know it's there? There is no way of seeing it directly. But there are ways in which we can confirm its presence. One of them is gravitational lensing. As explained by Albert Einstein, mass distorts the fabric of space-time. The heavier an object is, the more it will warp space-time. As light travels through this distorted space-time, its path gets diverted based on the object's mass. If there is a large cluster of galaxies between Earth and a distant galaxy, this cluster will act as a lens, bending the path of distant light toward us. Have you ever seen images released by NASA that are stretched and oddly shaped? Now you know why.

The same phenomenon has been found in the case of dark matter. Astronomers have found that light is being deflected in space with little to no mass present. And the culprit here is dark matter. Just like ordinary matter, the gravity of dark matter also distorts the fabric of space-time, causing the gravitational lensing effect where it shouldn't be. Through this effect, we can trace the presence of dark matter everywhere in our galaxy. It can also reveal the distribution and amount of dark matter around us.

At first, scientists thought that dark matter was just a new type of ordinary matter that does not emit light. But this idea does not hold any ground. Dark matter is unlike anything we have ever seen. It impacts the formation and spin of galaxies. It's possible that dark matter is passing through your body right now. You can't feel it, as it doesn't interact with ordinary matter. Dark matter could be an undiscovered particle, but it does not act like any particle we know. If it is a particle, one thing we know is that it can interact with us in the form of gravity. What the nature of this particle is, and what it is made up of, is a point of discussion.

Another possibility is that our current understanding of gravity is incomplete. Our current theories break down when we talk about the gravity of dark matter. Everything we know about dark matter comes from the understanding of its gravitational effects. By studying the radiation left over from the Big Bang explosion, we can identify where more radiation exists. More radiation means more matter, or that dark matter was created there. This way, we can locate dark matter across the universe.

Some scientists believe that dark matter is not from our world. It comes from a higher dimension or a higher world. But for some reason, the gravity of dark matter is leaking into our universe. If that is the case, it will prove that gravity can travel between multiple dimensions and open doors to new worlds.

Dark Energy

We know about the 5% of ordinary matter that we're all made of. Next, we have the 27% of the dark side of matter—dark matter. We cannot directly interact with it, but we know it's there. So, what is the remaining 68% of our universe made of?

At the beginning of the 19th century, astronomers believed that space was nothing but an empty void. It didn't have any properties of its own. At the same time, Albert Einstein was working on his Theory of Relativity. What he discovered changed the course of astronomy. While working on relativity, Albert Einstein discovered that space isn't just an empty void. Space itself can have its own properties. Einstein predicted that it is possible for more space to come into existence. Empty space can have its own energy.

Since this energy is a property of space itself, when more space comes into existence, more energy will be generated. At that time, the static model of the universe was widely accepted by the scientific community. Theoretically, Einstein determined that our universe must be expanding, but that idea did not sit well with other known cosmological facts at that time. So, to support a static model of the universe, Einstein dropped his idea of the expanding universe and added a new term to his equations: the cosmological constant.

In the late 1920s, a talented astronomer, Edwin Hubble, studied the deep universe. Distant galaxies, stars, and supernovae were his points of interest. One thing that interested him the most was supernova explosions. With their help, Hubble tried to figure out if our universe was expanding. After long observations and studying supernovae in multiple galaxies, Hubble concluded that our universe is not static at all and that, in fact, it is expanding. With this discovery, the static model of the universe was put to rest. This discovery also made Albert Einstein drop the cosmological constant from his equations.

The model of the universe changed with the discovery of an expanding universe. As the Big Bang theory gained more interest among scientists, the common belief was that this expansion was just an aftermath of the Big Bang. Since gravity has infinite attractive strength, the expansion must be slowing down. The gravitational force of all the objects in the universe must be working collectively to stop the expansion. Another question came for astronomers to answer: At what rate is our universe slowing down?

To find the answer, astronomers turned their telescopes toward exploding supernovae. What they discovered blew their minds. Instead of slowing down, the rate of expansion was increasing exponentially. There is something in the universe working actively against the force of gravity, driving this expansion. Scientists later named it dark energy. By calculating the amount of energy needed to overcome gravity and expand the universe exponentially, scientists concluded that dark energy makes up about 68 percent of the universe.

It is believed that, just like ordinary matter and dark matter, dark energy was also created after the Big Bang. It's been here since the beginning, and we've just figured it out recently. After the Big Bang, as the universe expanded rapidly, dark energy took over the force of gravity. With the expansion, the strength of dark energy also increased exponentially.

There is both a positive and a negative impact of dark energy dominating the universe. If dark energy were not as strong as it is, the collective gravitational force of matter and dark matter would have stopped the universe's expansion long ago. It would have reversed the process of expansion. As a result, our universe might have contracted, ending up as a gigantic ball of matter and energy. That would be a nightmare for all the species living on this planet. It's also possible that the universe might have contracted long before our solar system was formed. Thanks to the repulsive nature of dark energy, we live in a universe where galaxies and stars are not getting closer over time and smashing into each other.

The repulsive nature of dark energy has prevented the superhot fate of the universe. At the same time, it has opened a new possibility. As a result of the expansion, distant stars and galaxies are moving away from us. There may be a time when we will not be able to see other galaxies.

It is possible that our current understanding of gravity and how it works is incomplete. We know about the gravity of matter and dark matter. The force of dark energy could be just another part of gravity, but opposite in nature. It's also possible that dark energy is a property of space itself. It originates with the birth of space and further increases its expansion. But it is still a point of discussion when it comes to what dark energy really is.

Early Galaxies

The early universe was full of massive hydrogen and helium gas clouds. When these clouds came together, they formed the very first stars. It is not easy for stars to shine and bring us their light. Gravity plays a significant role in this process. The early stars started shining when hydrogen began to fuse into helium and other heavier elements due to the gravitational force and heat of the star. The gravitational force not only provided extra pressure but also raised the core temperature, making fusion much easier.

Early stars in our universe were colossal compared to the Sun because their formation took place in large hydrogen and helium clouds. Also, there was little to no competition between stars to capture more matter. Matter was abundant; they just had to take it. Due to their gravitational force, those stars added more and more matter and continued becoming denser and heavier. As they became larger, they added more matter to their pile, becoming unstable. Most of those early giants ended up in supernova explosions, creating heavier elements. The early universe was lit up, and explosions were everywhere. However, not all stars exploded. Some of them collapsed under their own gravity, resulting in the formation of black holes.

It was the best time for black holes to exist because the universe was small, dense, and full of matter for them to feed upon. So, the black holes started sucking more and more matter, becoming massive and supermassive black holes. Because of their enormous gravitational attraction, long-distance matter also found itself revolving around them. Because of the rotation, the disk shape around them turned spiral and concentrated toward the center. This structure remained stable for the next billions of years.

The surviving matter did not fall into the black hole due to its angular momentum. Neither did it break away because of the gravitational attraction, and it kept on steadily revolving around the black hole. As time passed, the revolving matter began to form its own stars, planets, and eventually solar systems. This complete structure is known as a galaxy. There is a supermassive black hole at the center of almost every galaxy in the universe.

Galaxies are being formed to this day; it has been found that nearly all the galaxies we see today were formed shortly after the Big Bang. Our Milky Way galaxy is considered to be roughly 13.6 billion years old. We are unsure whether the stars that formed first gathered into galaxies or the galactic clouds formed first, from which the first stars were born. It is possible that the galaxies were formed where the dark matter and ordinary matter clumped together due to the irregularities in the distribution left over from the Big Bang.

GN-z11

It is essential to talk about GN-z11 because it is one of the oldest, if not the oldest, galaxies in the universe. Identified in March 2016, GN-z11 is also the most distant galaxy in the observable universe. Scientists observed GN-z11 as it was 13.4 billion years ago, just 400 million years after the Big Bang. Due to the universe's expansion, the current distance of this galaxy is approximately 32 billion light-years.

GN-z11 belongs to the first stars and galaxies formed after the Big Bang. When this galaxy was born, the universe emerged from a period known as the Dark Ages. During this period, the entire universe was covered in darkness. Stars that make our universe shine did not exist. However, that era did not last too long, as new stars and galaxies began to form as soon as the universe cooled down.

GN-z11 does not have a massive size; it is about 25 times smaller and just 1% of the mass of our Milky Way galaxy. All galaxies form new stars at a specific rate, depending on the amount of gases and dust. But GN-z11 forms new stars at 20 times the rate of our Milky Way, proving that at the beginning, galaxies formed stars rather quickly. Since it is forming new stars quickly, it is very bright, making it possible for astronomers to detect it. Observing GN-z11 is a significant step back in time; we look at creation itself in its earliest form. When we observe GN-z11, we observe the very beginning of the universe, because light from such a distant galaxy travels vast distances to reach Earth. When we look at GN-z11, we are looking at it as it was 13.4 billion years ago.

Supernova

We are stardust. A supernova is an event that releases an enormous amount of energy in a short period of time. They occur when the life cycle of a massive star comes to an end. The majority of stars in the universe are average in size. When they are born, they light up the space around them, and after a few billion years, they become red giants. Their luminosity keeps decreasing with time, and eventually, they fade away. Before fading away, their temperature reaches such low levels that you can touch those stars with your bare hands.

Our Sun itself is an average-sized star and is one of them. It has been around for over 4.6 billion years and will remain here for the next few billion years. In about 110 million years, our Sun's luminosity will increase by over 1 percent. This increase will not be noticeable directly, but it could threaten life on Earth. In about 1.1 billion years from now, the Sun's luminosity will increase by 10 percent, causing the average temperature on Earth to reach over 45°C. This would be a severe threat to humans and all life on Earth. By this time, almost all species would have died on Earth. Earth's atmosphere will become a moist greenhouse, and our oceans will evaporate at an alarming rate.

In about 5.4 billion years, the Sun's hydrogen supply at the core will get exhausted, and no more fuel will remain for fusion to occur. As a result, the Sun will begin to evolve into a red giant and start expanding. Earth will receive more light and become hotter every day. Being the closest planet to the Sun, it will first consume Mercury and Venus. In about 7.6 billion years, the Sun will have expanded so much that it will likely consume our home planet.

It will contract quickly in about 8 billion years, becoming a white dwarf star. The Sun will lose over 50% of its current mass in this quick expansion and contraction process. White dwarf stars do not emit as much energy as regular stars. So, by this time, if our planet is not eaten by the Sun, its surface temperature will start to drop rapidly. The Sun will have cooled to five degrees above absolute zero in about one quadrillion years. It will be so cold that you would not be able to touch its surface with your bare hands. The Sun will not be shining at all. It will have become a black dwarf with no light emission. Its core would also have cooled down significantly, and fusion would stop completely.

All the stars we see in the night sky will use all of their energy and be gone one day. What happens before that depends upon their mass. Most of the stars in our universe are average-sized. Most of them will die like the Sun, but not all. Only the stars whose mass is 0 to 8 times that of the Sun will die like this. Stars whose mass is 8 to 20 times the mass of the Sun have a different fate.

Have you ever wondered why it takes millions or billions of years of fusion for a star to become unstable and collapse? Why do they not collapse into themselves as soon as they are made? After all, they are purely made up of gases like hydrogen and helium with no solid core. Well, that is due to fusion. The gravity of a star wants to collapse, but at the same time, it is being held in place by the outward force of fusion. The more gravitational force pressurizes the core, the more fusion occurs. This is why bigger stars die early; they burn matter much more quickly than smaller stars.

A star is an equilibrium of its gravitational and nuclear forces. The gravitational force is caused by its mass, and the fusion occurs due to nuclear forces. When stars use all of their fuel, and no more fuel is left to burn, this equilibrium is disturbed. Stars whose mass is 8 to 20 times the mass of our Sun collapse into themselves due to this pressure difference. Boom. A supernova explosion occurs. The bigger the star is, the bigger the supernova explosion. These stars light up the entire galaxy with their explosion. The effects of their explosion can be seen in nearby galaxies and can be detected thousands of light-years away.

These stars create the building blocks necessary for life with their explosive fate. Our element factory (stars) cannot form heavier elements like iron, gold, etc., on a large scale through fusion. Their core temperature can only fuse hydrogen into helium and some sort of lithium in massive stars. So, if we want to create large amounts of heavier elements, we need very high temperature and pressure. Nothing can bring this temperature except when a star collapses into itself and blows up in the form of a supernova. A supernova explosion provides enough energy to form heavier elements. The inner temperature of the star reaches millions of degrees, enabling nuclei to fuse into each other. We have iron in our blood, which came from a supernova.

Heavy elements are not the only things that these stars leave behind. Neutron stars are also born from these explosions. When giant stars die in the form of a supernova, they crush protons and electrons into neutrons at the core. So eventually, what is left is a massive ball of neutrons that we call neutron stars. Any denser than that, they would become black holes. Neutron stars are typically tiny but extremely heavy. They have a radius of roughly 12 kilometers, but their mass could be 1.4 to 2.2 times that of the Sun.

Neutron stars are the tiniest and densest stars ever known to exist. Their density can be imagined from the fact that one teaspoon of neutron star material would weigh roughly 10 million tons. Neutron stars also rotate very fast; they rotate up to 43,000 times per minute without falling apart. It has been predicted that our Milky Way galaxy alone hosts about 100 million neutron stars. Most of the neutron stars we have observed are extremely hot. Their surface temperature can reach 60,000 K, much more than that of the Sun, which is 6,000 K.

Pulsars are also born the same way. As we know, almost all stars rotate about their axis. However, when a star goes supernova, it loses a lot of its mass. So, to maintain its angular momentum, the remaining star, the neutron star, must spin faster. When a neutron star spins that fast, it blasts small radiation beams. When we observe such stars from Earth, we see pulsating light beams. So instead of calling them average neutron stars, we call them pulsars.

Stars that are heavier than 20 times the mass of our Sun have a different fate. Whenever these stars run out of fuel, instead of collapsing into themselves and then exploding into a supernova, they only collapse into themselves, forming a black hole. We will talk about them later in detail.

A supernova is one of the most extreme events in the universe. When we talk about a supernova, we talk about a star exploding into bits and pieces in a small fraction of a second. An entire star could collapse, creating either neutron stars or black holes. They release more energy in a fraction of a second than our Sun will release in millions of years. Supernovas are characterized into two different categories, i.e., Type-1 and Type-2. This distinction is generally based upon their dramatic way of explosion and the type of star involved.

Supernovae are rare; they are not easy to detect even in our galaxy. Since the invention of the telescope, we have been able to observe only around ten supernova explosions in our galaxy. Astronomer Johannes Kepler, in 1604, observed one of the first supernovae ever. He observed this supernova even before the invention of the telescope, with the naked eye. However, calculations show that there should be a minimum of three of these explosions occurring in a century in our Milky Way galaxy.

Betelgeuse

Betelgeuse is one of the shiniest stars in the night sky. When you are stargazing from your roof, one of the stars you see in the sky might be Betelgeuse. It is important to learn about Betelgeuse because it can go supernova anytime soon. Betelgeuse is a red supergiant star located 642 light-years away from Earth. This massive star has a diameter of about 1.2 billion kilometers. Betelgeuse is very young, roughly 10 million years old, much younger than our Sun. But due to its immense size, it is burning its fuel very rapidly. Some astronomers suspect that this star may go supernova within 100,000 years. One million years is the maximum estimated time until this red supergiant star explodes.

When Betelgeuse explodes, it will light up our sky for several months. Its light will be so bright that it will be visible during the daytime. Recent high-resolution images show that Betelgeuse is going through some internal changes; we do not know what they are. It is not clear, but it is possible that Betelgeuse has already exploded, and we do not have to wait 100,000 years to see this supernova. But distance is the problem. Betelgeuse is located 642 light-years away, which means light will take 642 years to reach our planet after it explodes. All we can do is wait and watch the sky.

Black Holes

John Michell was the first person to suggest the presence of black holes, which he called "dark stars." He predicted that when a body is so dense that its escape velocity is close to the speed of light, it will turn into a black hole. He said that we could only observe them by their gravitational effects. Scientists speak in a different tone when it comes to black holes. For some, they could be the pathways to new dimensions. For others, black holes are nothing but ultra-dense space that does not let anything out.

Albert Einstein explained that black holes curve the space-time fabric more than anything else because of their infinite density. As light travels through that curved space, it bends forever and can never come out of that region. It is said that black holes are not black but instead are the brightest objects in the universe; they just do not allow the light to reflect. Our Milky Way galaxy contains more than 100 billion stars. If we assume that 1 out of every 100 stars has enough mass to end up as a black hole, then there are over 100 million potential black holes in our galaxy alone. There are enough black holes in the universe, but we don't see them because they don't let us. Almost all galaxies have a host black hole located at the center. The Milky Way also has a supermassive black hole at its center known as Sagittarius A*.

In general, a black hole is a place in space where the gravitational force is so strong that not even light can escape. Black holes come in different shapes and sizes, depending on the amount of matter they have sucked in and the star's mass that resulted in their formation. There is no specific size for black holes. There are black holes that are less than 25 kilometers in diameter, while others are supermassive, with diameters of billions of kilometers.

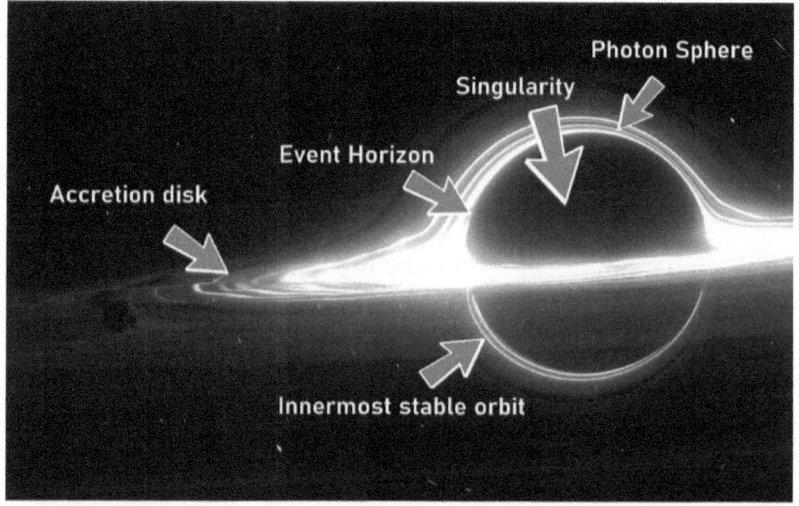

To understand a black hole better, let us try and get into one. Suppose you have a spaceship that can take you to a black hole. Let us find out what will happen as you approach the black hole. In reality, you will get stretched like spaghetti, your body will break down into atoms, and you will die long before reaching it, in a process called spaghettification. However, we are assuming that the black hole's gravity has no impact on us in this case.

Accretion Disk

You will first encounter the accretion disk as you get close to a black hole. An accretion disk is a disc of superheated gases and dust swirling around the black hole at a very high speed. This superheated gas produces electromagnetic radiation (such as X-rays) that we generally use to locate a black hole. Matter from the accretion disk falls into the black hole, giving it more strength. It is the "lunch" of a black hole, which gives it more energy.

Innermost Stable Orbit

As we pass through the accretion disk, the next place we will encounter is the innermost stable orbit. This orbit is the inner edge of the accretion disk. It is the last place from where we can turn back and come out safely. Once we cross this point, there is no going back. In simple words, it is the last stable circular orbit with a minimum radius for a particle to revolve around the black hole.

Photon Sphere

Before falling into the black hole, we will encounter the photon sphere. The photon sphere is a spherical region around the black hole where gravity is so strong that even photons (light particles) are forced to travel in an orbit. This means that the black hole's gravity bends their path, and thus, they are forced to orbit the black hole until they fall into it or spiral out into space.

Event Horizon

The event horizon is the point of no return. Once you cross the event horizon, you are forever stuck inside the black hole, and there is no way out. The event horizon is the radius around the singularity. It is the boundary that separates a black hole from the rest of space. The escape velocity within the event horizon exceeds the speed of light. You must travel faster than the speed of light to get out of the event horizon. Physics does not allow such travel. Going forward is the only thing left.

Singularity

After crossing the event horizon, we come to the singularity. It lies at the very center of a black hole. The singularity is where matter has collapsed into a point of infinite density. All the matter that falls into a black hole eventually ends up here. We are not sure about the nature of the singularity and what it is. The singularity is still a point of discussion among scientists. The laws of physics we know break down at this point, making them hard to understand.

Hawking Radiation

Physicist Stephen Hawking said that black holes might look stable, but they are not. They evaporate in the form of Hawking radiation. The smaller the black hole, the faster it evaporates. Hawking explains it in terms of space and how it works. Free space does not mean nothing; it consists of particles and antiparticles that come into existence, and soon after, they annihilate each other. This process occurs continuously and everywhere in space. The process of creation and annihilation of particles occurs near the black hole as well.

Something crazy happens at the event horizon. Hawking said that as soon as particles and antiparticles form at the event horizon, one gets sucked into the black hole while the other particle stays out. The other particle may escape and have no one to pair with. The particles come out in the form of radiation that we call Hawking radiation. This phenomenon reduces the life of a black hole. However, this process is relatively slow. Given enough time, the black hole will radiate away its mass and eventually vanish. But if the black hole continuously devours more matter, it will stay.

Based on their mass and how they are formed, black holes are divided into four categories: stellar, intermediate, supermassive, and miniature black holes.

Stellar Black Holes

Stellar black holes are the most common black holes. These black holes are formed by the death of stars. As stars reach the end of their lives, most of them inflate, lose mass, and eventually become white dwarfs. However, stars whose mass is more than 20 times the mass of our Sun become a black hole. These black holes are known as stellar black holes. Stellar black holes can be found almost everywhere in the universe. Their mass is generally between 5 and several tens of solar masses.

Intermediate Black Holes

Intermediate black holes are a particular class of black holes. These black holes have a mass of 100 to 100,000 times the mass of our Sun. Intermediate black holes are massive compared to stellar black holes, but smaller than supermassive black holes. Several intermediate-mass black holes have been found in our Milky Way galaxy. Scientists have been able to trace their presence by observing the gas clouds and accretion disks around them.

Supermassive Black Holes

As their name suggests, supermassive black holes are heavy. They are one of the biggest objects in the universe, only behind galaxies themselves. Their mass is in the order of hundreds of thousands to billions of times the mass of the Sun. Astronomers believe that supermassive black holes are formed from the collapse of massive gas clouds during the early stages of the formation of a galaxy. As a result, supermassive black holes are generally found at the center of most galaxies. The bigger the galaxy is, the bigger its central black hole must be.

Sagittarius A* is located at the center of our galaxy, roughly 26,000 light-years from Earth. It has a diameter of 44 million kilometers and is 41 million times the mass of our Sun. Sagittarius A* is also a powerful radio source, giving off strong radio waves, likely originating from the matter orbiting around it. Astronomers have not seen Sagittarius A* with a telescope. Instead, they have noticed the strange motion of stars around it. That can be explained only by the presence of a massive object at the center.

Miniature Black Holes

Miniature black holes are hypothetical tiny black holes. They have a very low mass. The concept of miniature black holes was introduced in 1971 by Stephen Hawking. The problem with miniature black holes is their small size. Even if they are formed, they would radiate away almost instantly. Some scientists predict that miniature black holes can be formed due to the high energies available in particle accelerators, such as the Large Hadron Collider.

First Image of a Black Hole

Previously, it was thought impossible to capture the image of a black hole since they do not allow light to escape. However, with advancements in technology, scientists have made it possible. We have studied black holes for a long time, but none of us had seen one until recently. Messier 87 is an elliptical galaxy located in the constellation Virgo. It is one of the biggest galaxies observed in the universe. It is located about 53 million light-years away from Earth.

At the heart of the Messier 87 galaxy, there is a supermassive black hole, designated M87*, with a mass 6.5 billion times that of the Sun. Astronomers used the Event Horizon Telescope to study this black hole based on the data gathered from this source. They created the first image. This image contains a rotating disk of ionized gas surrounding the black hole. Matter continuously falls into the black hole and keeps feeding it. The disk rotates at a very high speed of roughly 1,000 km/s.

Until now, we had only seen indirect evidence of black holes. We were able to study the high-energy jets shooting straight from them. We were able to study the X-rays coming from the matter circling the black hole. Not just X-rays, but we were also able to detect the gravitational waves supposed to be coming out of colliding black holes. However, this image of a black hole taken in 2019 is one of the most significant landmarks in human history.

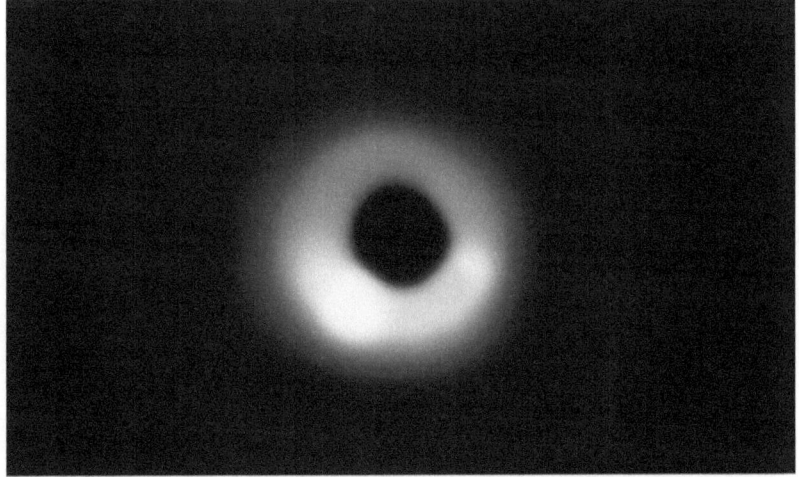

If black holes are black, how do we know they are there? A black hole cannot be seen directly, but scientists can observe the effects of the black hole's strong gravity on the stars and gases orbiting it. If a star is orbiting a point in space and it is not clear what it is orbiting around, we can study its motion to see if that point is a black hole. Generally, high-energy light is produced when a star and a black hole are orbiting very closely.

A black hole's gravity can sometimes steal the outer gases of a star. However, that gas does not directly fall into it; some does, but most of it orbits around the black hole in the accretion disk. While orbiting, it gets heated to a point where it starts releasing X-ray light in all directions. Space telescopes can measure this light.

Our Solar System

A solar system is an arrangement of one or more planets around a star. The Sun is our star and the planet we live on is Earth. The journey of our solar system began more than 5 billion years ago with a massive cloud of dust and gas, mainly containing hydrogen and helium. A new star emerged from the very center of that gigantic cloud. That was our Sun. Five billion years ago, there was no sign of a planet, let alone a blue planet.

However, as the dust cleared, gravity began to force bits of matter to clump together, forming a large number of planets. There was a beautiful structure of planets around this newly born star. Scientists believe there were hundreds of planets in our solar system initially, but not all had a stable orbit. Some of them fell into the Sun, becoming a part of it. At the same time, others left the orbit and flew away into the darkness of the Milky Way galaxy. Many of them collided, forming giant planets like Jupiter and Saturn.

With every impact, planets grew. Planets that moved faster and had large orbits grew bigger because they swept more matter while revolving around the Sun. Jupiter is an excellent example of that. Mars could have been a massive planet, but it could not get enough matter in its orbit. It is hard to believe that dust and gas clouds combined, forming planets in a process that took millions of years—not to mention that Earth was made in the same way.

Planets around the Sun are divided into two categories: inner and outer planets. The inner planets include Mercury, Venus, Earth, and Mars. All of them are primarily composed of rocks and metals. In the beginning, these planets were nothing but hot balls of lava whose surface temperatures were thousands of degrees Celsius. These planets had an atmosphere full of CO_2, nitrogen, and other harmful gases. The planets were boiling spheres where volcanic activity was widespread. It was a living hell on the surface.

The outer planets include two gas giants, Jupiter and Saturn, and two ice giants, Uranus and Neptune. The gas giants are primarily made of gases like hydrogen and helium. These planets are also referred to as failed stars because of their similarities in composition to the Sun. If Jupiter were 13 times its current mass, it would have become a brown dwarf, and there would be some fusion occurring at its core. However, to become a small red dwarf star, it would need to be 80 times heavier.

The ice giants are primarily made of dense icy materials such as water, ammonia, and methane. These planets also have a small rocky core at the very center. If you were to stand on the surface of these planets, you would descend into thick clouds until you reach the core.

The Sun

The Sun is our host star. It dominates the solar system with around 99.8% of the total mass. If we compare our central star with Earth, it has 109 times the radius and around 333,000 times the mass of Earth. The escape velocity from the outer surface of the Sun is also very high—around 617 km/sec, about 55 times that of Earth. When it comes to our place in the galaxy, our solar system is an average of 27,000 light-years away from the center of the Milky Way. The Sun orbits at

~230 km/sec around the center of the Milky Way. We do not notice this movement because we live in Earth's protective atmosphere.

Nearly three-fourths of the Sun is pure hydrogen, whereas one-fourth is helium. There is also a negligible amount of heavy elements such as oxygen, carbon, iron, neon, etc. Scientists estimate that the core of our Sun is over 15 million degrees Celsius. This is where the process of converting light elements into heavier elements occurs. Being an average-sized star, the rate of fusion in the Sun is slow. Today's Sun is likely the same as it was 4 billion years ago and will remain the same for another 2–3 billion years, except that its luminosity will increase slightly over time.

The distance between the Sun and Earth is used as a parameter to measure large cosmic distances—also called an astronomical unit or AU. The average distance between the Sun and Earth is about 150,000,000 km or 93 million miles. One astronomical unit equals 149,597,870 km. We have a stable future because we have a stable host star nurturing life on Earth with its light. All plant life uses this light for photosynthesis and gives us the oxygen we breathe. All the sunlight we receive is just a tiny fraction of the Sun's total energy output. Yet, if we could capture only 90 minutes of sunlight, it would be enough to power the entire world for a full year.

Our Sun is essential because its heat keeps us alive and its energy enables plants and trees to feed. It's crazy to think, but our planet is rotating at roughly 1,600 kilometers per hour. It is also revolving around the Sun at about 108,000 kilometers per hour. Our Sun is orbiting the center of the Milky Way at 828,000 kilometers per hour. At the same time, our galaxy is moving at 600 km per second in the universe with respect to extragalactic frames. All of this is happening as you read this book.

Mercury

Mercury is the first planet from the Sun and the smallest planet in our solar system. Mercury has a rocky surface with a large number of craters on it, much like the dark side of the Moon. These craters show the brutal bombardment that occurred for billions of years after the formation of Mercury. Mercury is one of the two planets in our solar system that does not have a moon. It has a thin atmosphere consisting of hydrogen, helium, oxygen, sodium, calcium, potassium, and water vapor. Being the closest planet to our Sun, its surface temperature can reach above 420 degrees Celsius.

Mercury is tidally locked with the Sun in a 3:2 spin-orbit resonance. Relative to the Sun, it rotates on its axis three times for every two revolutions around the Sun. The side of Mercury that faces the Sun has a very high temperature, whereas the opposite dark side has freezing temperatures. Due to this dramatic temperature variation and toxic atmosphere, there are no chances of survival on this planet. Mercury takes only 88 Earth days to complete one revolution around the Sun, making it the fastest planet. It has a surface density of 5.51 g/cm^3, slightly lower than Earth, making it the second most dense planet.

Venus

The next planet is Venus. Venus is very similar in physical characteristics—such as size, mass, gravity, etc.—to Earth. Due to this fact, Venus is also called Earth's sister planet. The inner core conditions of Venus are also similar to Earth. Venus has an atmosphere consisting mainly of carbon dioxide—more precisely, over 96% carbon dioxide and about 3.5% nitrogen. This high carbon dioxide density makes Venus a perfect example of the greenhouse effect. Being closer to the Sun, Venus receives much more sunlight than Earth. Most of the sunlight gets trapped by the carbon dioxide in its atmosphere and cannot leave. This raises the overall temperature, reaching up to 460 degrees Celsius, making it the hottest planet around the Sun, hotter than Mercury.

Venus has sulfuric acid clouds, so whenever it rains, sulfuric acid drops from the sky—which is a horrifying scene to imagine. The presence of carbon dioxide in its atmosphere and the high temperature make it unsuitable for life. Venus has a thicker atmosphere than Earth, resulting in a lethal surface pressure 92 times that of Earth. Venus takes 224.7 days to complete a single revolution around the Sun. After the Moon, it is the second brightest natural object in the night sky. It looks like a small bright spot when observed with the naked eye.

Earth

The third planet from the Sun is our home planet, Earth. Just look down at the ground—that is Earth. Earth is neither too hot nor too cold; it is perfect for the growth and well-being of life. Earth is the only planet we know that supports life. We will briefly discuss this planet in a later chapter.

Mars

The fourth planet is Mars, the red planet and the next home for human civilization. The abundance of iron oxide in its atmosphere makes it reddish. Mars is smaller than Venus and

Earth, but it is larger than Mercury. Mars has a fragile atmosphere consisting of CO_2 and oxygen. The presence of water ice over a large part of Mars makes it an exciting planet that could support life in the future. The discovery of water in the form of ice on Mars indicates that this planet may have supported life in its past. However, Mars lost its atmosphere and its oceans of water due to solar storms and a lack of a magnetic shield. Mars is the next target for many space research organizations like NASA and SpaceX to land humans on its surface. Humans are yet to land on its surface, but our rovers are already there. The Curiosity rover is one of them.

After Earth, Mars is the next planet that has its own moons. The beauty of Mars is enhanced by its two moons, Phobos and Deimos. Its moons are not as big as our Moon, nor are they round like ours, but they are beautiful. Because of their dramatic structure, scientists believe that these two moons might be large asteroids captured by Mars's gravity. They might have arrived from the asteroid belt between Mars and Jupiter.

Asteroid Belt

The asteroid belt is located between our solar system's inner and outer planets, i.e., Mars and Jupiter. It contains a large number of asteroids orbiting the Sun. Hundreds of thousands of asteroids are found in the asteroid belt, but almost half of its mass is comprised of four bigger asteroids. These are Ceres, Vesta, Pallas, and Hygiea. Among these asteroids, Ceres is also designated as a dwarf planet. This means that it is neither too small to be a proper asteroid nor too big to be considered a planet. Ceres has a diameter of about 946 kilometers. The asteroid belt consists of various solid asteroids, comets, and irregularly shaped bodies, some of which are as small as a particle, whereas others are over 900 kilometers wide.

There are two leading theories that describe the presence of the asteroid belt. The first theory is that it is just as it was at the beginning of the solar system. When our solar system formed, gas and dust combined, forming small asteroids and comets, but they could not form a proper planet. If we assembled all the asteroid belt's mass, it would make an excellent small-sized planet. But since the asteroids could not combine, they have been here since the beginning. Now and then, we observe asteroids and comets leaving the asteroid belt due to the influence of the gravitational force of either Mars, Jupiter, or even other objects within it. Those asteroids and comets often fall into the Sun or leave the solar system. We never know when the next asteroid or comet will leave the belt and approach our planet.

The second theory is that there were two small-sized planets between Mars and Jupiter in the early solar system. But somehow, their orbits intersected with each other. As a result, a massive collision between them occurred, forming the asteroid belt. A large amount of debris either fell into the Sun or escaped the solar system, and only a few hundred thousand objects remained in the belt.

Jupiter

The fifth planet from the Sun is Jupiter. Jupiter is the biggest planet and the second biggest object in our solar system after the Sun itself. Jupiter has no solid surface. It is mainly made up of gases. Since there is no land on Jupiter, if you try to land on it, you will get sucked toward the center. Some scientists believe that Jupiter might have a solid core made up of metals because of its tremendous pressure. However, we have no substantial evidence supporting this prediction because we cannot go there and observe it directly. It is all gas as far below as we can see. Jupiter has nearly 79 known moons. Most of these moons are very small, but some are the largest moons in our solar system. With a diameter of 5,262 kilometers, Ganymede is one of Jupiter's 79 moons; it is even bigger than the planet Mercury. If Ganymede were not orbiting Jupiter, it would be a planet by itself.

Jupiter is not as cool and calm as we see it in the pictures from Earth. There are terrible storms on Jupiter. Most of them are bigger than the biggest storms seen on Earth. One of its storms can be seen from the surface of Earth. It is known as the Great Red Spot. This storm has been going on for the last 200 years. It is so violent that the wind speeds can reach over 430 kilometers per hour.

Saturn

After Jupiter comes Saturn, another gas giant. Saturn is the most beautiful planet to observe through a telescope. It is also the farthest planet visible from Earth with the naked eye. Saturn is famous for its beautiful rings. These rings were once thought to be solid, but they are actually made up of rocks, ice, and other stardust. It is still not clear how old Saturn's rings are. Saturn's rings may have been around for 4.5 billion years, since the beginning of our solar system. Another possibility is that Saturn acquired its rings recently through the collision of two or more of its moons. Icy moons such as Enceladus have been found orbiting Saturn, so it will not be surprising if Saturn got its rings by ripping apart one of its moons with the force of its own gravity.

Saturn is the least dense planet in the solar system, with a density of 0.68 g/m³. If you put Saturn in a giant tub of water, it would float. Saturn has more than 60 known moons, and some are the same size as Mercury. Another thing that makes Saturn interesting is Enceladus. Enceladus is a small ocean world fully covered in ice. We cannot see what is below the ice sheet, but scientists have predicted that there could be vast oceans of water beneath the ice sheet. Scientists are looking forward to sending underwater drones to Enceladus to check if there is any underwater aquatic life.

Uranus

The seventh planet from the Sun is Uranus. Uranus is much smaller than Jupiter and Saturn. Uranus has small rings around it, much smaller than those of Saturn. We usually do not see them in the images coming from NASA. It is the coldest planet in our solar system, where the lowest temperature can reach -224.2 degrees Celsius. Since Uranus is so distant from the Sun, it takes a very long time to complete one orbit around the Sun—roughly 84 Earth years. It is the second least dense planet, with an average density of 0.687 g/cm^3.

Uranus's upper atmosphere is covered by methane gas, which gives it a beautiful blue color. We do not have much information about this gas giant because it has been visited only once by Voyager 2. This visit, for the first time, revealed the secrets of planets located in the outer solar system. All the planets in the solar system rotate on their axes; some rotate slowly, and some rotate quickly. In many cases, such as Earth and Mars, they have an axial tilt. Earth is tilted about 23.5 degrees away from the Sun, while Mars is tilted 24 degrees. This tilt results in seasonal change. Uranus is different; it has an axial tilt of almost 98 degrees. In a way, Uranus is like a ball rolling around the Sun in a circular pattern.

Neptune

The eighth and most distant planet from the Sun is Neptune. Neptune is another ice giant with a similar composition to Uranus. It is mostly made of icy materials such as water, ammonia, and methane. Neptune has 14 known moons. The average distance between Neptune and the Sun is around 30 astronomical units. It takes a very long time to go around the Sun—close to 164.8 Earth years. Neptune is famous for its strong winds that can reach well over 2,000 kilometers per hour. Neptune is also among the coldest planets in the solar system. In its top clouds, the temperature is well below −220 degrees Celsius.

Neptune does not have a surface that you could stand upon, but if you could, you would feel something amazing. The gravity of Neptune is almost the same as that of Earth. Neptune is 17 times heavier than Earth, but also 4 times bigger. Overall, its gravitational force is only 17% higher than Earth's gravity, so you would not notice any significant difference in that regard. Voyager 2 visited Neptune back in 1989, passing within 3,000 kilometers of the planet's north pole. Just like Uranus, Neptune also has rings, but they are very small.

Pluto

Discovered in 1930, Pluto was once listed as the ninth planet of our solar system. However, Pluto is too tiny to be considered a planet—it's even smaller than many moons we know today. In 2005, Eris, which is much bigger than Pluto, was discovered. When people started discovering other objects larger than Pluto and demanded their object be granted the status of a planet, Pluto was removed from the list of planets.

In 2006, after 76 years of being listed as a planet, Pluto was declared a dwarf planet, which means that it is like a planet but too small to be called a planet. There are at least five known official dwarf planets in the solar system discovered after Pluto, and there could be many more. There are estimates of up to 200 possible dwarf planets in the Kuiper Belt.

Kuiper Belt

The Kuiper Belt is a circumstellar disc located in the outer Solar System. It is similar to the asteroid belt, but it is much more massive and broader. Where Neptune's orbit ends, the Kuiper Belt starts. It extends to a distance ranging from 30 to approximately 50 AU from the Sun. For us, this place is not only mysterious but also very cold and dark. Since this region is so far away, the Sun's heat and light do not reach there effectively. As a result, the Kuiper Belt is thought to be the source of comets we receive in our solar system. So far, we have mapped only about 2,000 objects from this region, but there are many more.

Astronomers believe that the Kuiper Belt is the remnant of our solar system's formation. The Kuiper Belt might have formed a planet if Neptune did not exist. However, Neptune's gravity has stirred up this region and prevented the objects from forming a planet. The Kuiper Belt is one of the largest structures in the solar system, second only to the Oort Cloud.

Oort Cloud

The Oort Cloud is a shell of icy objects located in the outermost region of our solar system. The Oort Cloud encloses the Sun at distances ranging from 2,000 to 200,000 astronomical units. Some astronomers believe that the Oort Cloud has not existed forever. Instead, the Sun might have captured the material in the Oort Cloud from the outer disks of other stars in our nebula.

It is hard to study the Oort Cloud because of its distance from the Sun. However, it is estimated to contain more than 2 trillion icy objects, comets, and asteroids. In a way, the Oort Cloud completes the solar system. It gives us a boundary beyond which we can look and say that this is how big our solar system is. Since it is too far and complicated, astronomers are yet to study most of its bigger objects and predict their trajectories around the Sun.

PART-III

Our Planet

Earth's Story

Every planet in the solar system has its own story. To understand Earth's story, we must follow the tracks left behind since its formation. Even though the early solar system's dramatic conditions have destroyed almost all the evidence, a few pieces still remain. One of them is the rocks that were formed billions of years ago. Rocks can reveal all the transitions our planet has been through. Geologists hunt for such rocks to reveal what Earth looked like back then. Rocks from that time are very hard to find today, so geologists generally use meteoroids and date them to get Earth's age. Some meteoroids found on Earth can be over 4.5 billion years old.

Our planet began forming about 4.5 billion years ago. In the beginning, it was in a molten state. There was nothing but large oceans of lava. If you stood on Earth, you would get submerged in lava and die instantly due to the intense heat. At the same time, Earth was getting bombarded by millions of large asteroids, comets, and meteoroids every day. This bombardment prevented the Earth's surface from cooling down quickly. In the molten state, most of the heavier elements, such as iron, nickel, and gold, moved towards the center. Today, when we study the center of our planet, that's where most of the heavy elements are. Lighter elements such as clay and sand stayed at the top of the surface. There was no sign of life, as the average temperature was over 1,000 degrees Celsius.

The vacuum of space is cold; its temperature is near absolute zero, –273.15 degrees Celsius. As soon as our planet started cooling down, it began to form the outer surface—the same surface that we stand on today. It took more than a million years for Earth to become cold enough to cover its surface with solid lava. Earth changed slowly. After millions of years of cooling down, Earth began to take shape and form. The lava oceans turned into hard surfaces—surfaces that can hold solid objects atop. From the outside, Earth looked like a dark and calm planet, but something else was happening inside. A lot of volcanic activity was about to take place. Over the next several million years, a large number of volcanoes erupted. All the heat and gases building up from Earth's formation suddenly spewed into the atmosphere, sending tons of dust, gases, and mainly CO_2 into the atmosphere.

Even after 100 million years of its formation, our planet was not suitable to support life. Earth had a solid surface, and volcanic activity was also slowing down with time, but there were two more ingredients that our planet needed to harbor life: water and oxygen.

How did Earth get all this water? Scientists still wonder how Earth acquired its water, but some believe they have found an answer. The answer to this question lies in the meteoroids. When scientists studied meteoroids falling on Earth from space, they discovered tiny liquid water droplets inside them. It is believed that as the universe cooled down, the water present in meteoroids became ice and stayed in outer space. When our planet was formed, meteoroids that struck Earth brought water with them. Millions of meteoroids collided with our planet every day, adding more water—drop by drop. Every drop of water on Earth is billions of years old. The water we drink may have traveled billions of kilometers in space inside a meteoroid.

It seems impossible to believe that small meteoroids can bring so much water to our planet and fill a large ocean. However, it is possible when the bombardment occurs for millions of years. We are lucky that the water did not cover the entire planet. Three-fourths of our planet is underwater, whereas the rest is solid land where we build our homes and live with our families.

If too many meteoroids had hit the Earth, covering the entire planet with water, only aquatic life would be thriving.

After 700 million years of Earth's formation, life-giving water covered its surface. The lava that erupted over the oceans cooled down quickly, resulting in the formation of small islands. In the future, these islands would join to form continents. Significant volcanic activity filled Earth's atmosphere with carbon dioxide. Nitrogen gas makes up 78 percent of the air we breathe. It is thought that most of this nitrogen was trapped in the primordial material that formed the Earth. When these materials collided, nitrogen was released. Still, there is one essential ingredient needed for life to thrive—oxygen. Where did the oxygen come from?

Every minute, we inhale and exhale several gases about 20 times. About 21% of that gas is oxygen. Early Earth had none of the oxygen we inhale today. The atmosphere was poisonous for life. Substantial volcanic activity filled Earth's environment with a mixture of methane, carbon dioxide, nitrogen, and even some sulfuric acid. We would not stand a chance in such a hostile environment. Stromatolites made it possible for Earth to have such a large amount of oxygen in an environment with nothing but poisonous gases. Stromatolites are found underwater; they contain microbes called cyanobacteria. Even a tiny piece of stromatolite can have millions of cyanobacteria. What makes these cyanobacteria unique is that they can produce oxygen.

Cyanobacteria changed the course of this planet with their ability to produce oxygen. Cyanobacteria take water and sunlight to produce oxygen. The process through which cyanobacteria produce oxygen is slow. It took millions of years of continuous pumping to fill Earth's atmosphere with oxygen. If we go close to cyanobacteria underwater, we can see them forming oxygen bubbles and releasing them. Life evolved and grew under the atmosphere that cyanobacteria created. Stromatolites can be seen even today in the depths of the oceans and in places where the water has evaporated. Living stromatolites are very rare to find. Stromatolites are an example of how microbial life dominated early Earth. Without stromatolites and cyanobacteria, Earth's geology would be very different. Water and a toxic environment would be there, and you and I would not exist.

The Moon

If we look at the entire human history, one object that has had the most profound impact on human lives is the Moon. Earth is unique because it has a special moon—the brightest object in the sky after the Sun. Since its birth about 4.5 billion years ago, the Moon has influenced Earth and the lives of its beings.

Earth spins at about 1,000 miles an hour at the equator; this number was much higher in the beginning. Our planet was spinning faster, and the Moon was much closer to the surface. Scientists believe that Earth and the Moon were formed at almost the same time. At that time, the Moon was orbiting only 32,000 kilometers from Earth, compared to today's 384,400 kilometers. We see tides in our oceans due to the Moon's gravity. When the Moon was much closer, and Earth was spinning faster, tides were much higher—as far as 100 meters or more in height. With time, Earth's rotation has slowed down, the Moon has moved away from us, and the waves are calm. Currently, the rate at which the Moon is moving away from Earth is roughly 3.82 cm per year.

The Moon is tidally locked with Earth, showing only one side to us. The force of gravity between Earth and the Moon causes some fascinating effects. The most obvious is the tides. The Moon's gravitational attraction is maximum on the side of Earth closer to the Moon and minimum on the opposite side. This effect of the Moon's gravity can be seen over the seas of Earth. Not being perfectly rigid, the Earth's oceans are stretched out along the line toward the Moon. In the same way, the Sun's gravity also plays a role, and as a result, we see two little bulges—one toward the Moon and one directly opposite, toward the Sun. This gravitational effect is much stronger on ocean water than on the solid crust. As the Earth rotates, these bulges move around the Earth once a day, giving rise to two high tides per day.

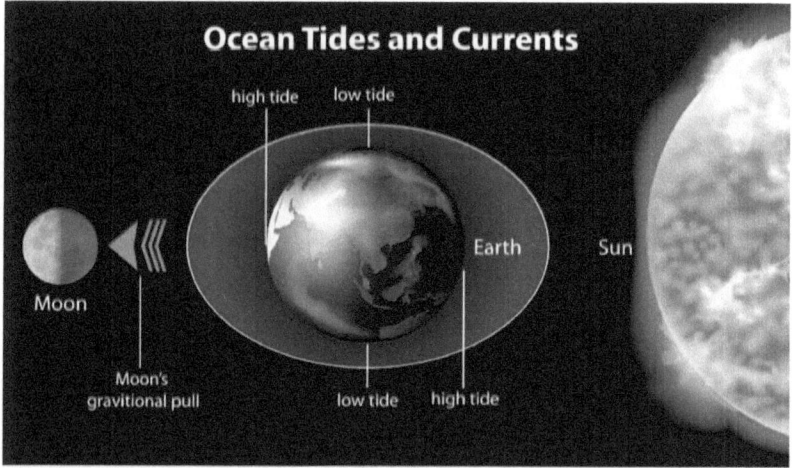

How the Moon came to be is still a mystery. But there are some theories that explain its existence.

1. Sister Theory

2. The Capture Theory

3. The Daughter or Fission Theory

4. The Impact Theory

Sister Theory

The Sister Theory suggests that during the formation of our solar system, the Moon formed as a separate object near Earth. The formation of Earth and the Moon took place at the same time. The material that formed Earth also gave birth to the Moon. As a result, in the beginning, we had two planets—one bigger (Earth) and the other smaller (Moon). It was a "double planet system."

However, there was a problem with this system. Earth took in more material from the raw disk of the solar system and acquired more mass, whereas the Moon could not. Earth was bigger; it had more gravitational force. So, Earth's gravity attracted the Moon, and it fell into orbit.

The Flaw in the Sister Theory

Even though this theory seems very satisfying, there is a significant flaw. The problem revolves around the density and composition of Earth and the Moon. When scientists studied the density of Earth and the Moon, it was found to be significantly different. Earth's density is around 5.5 g/cm^3, whereas the Moon's density is around 3.3 g/cm^3. If Earth and the Moon had formed as a double planet system side by side from the same interplanetary material, their densities should be the same. The difference in density proves that they did not originate together.

Capture Theory

As we know, Earth and the Moon's densities are not similar, which means that the Moon did not form anywhere near Earth. This is precisely what the Capture Theory suggests. According to this theory, the Moon was formed far away from Earth, somewhere outside our planet's orbit. However, it did not have a stable orbit around the Sun. As a result, the Sun's gravity pulled the Moon toward itself, and it started falling toward the Sun. As it passed near Earth, it got captured. Since then, the Moon has been orbiting in a stable orbit. If there were no Earth, the Moon would have either fallen into the Sun or collided with Mercury or Venus.

The Flaw in the Capture Theory

The main problem with the Capture Theory is the unusually large mass of our Moon. For us, Earth might seem like a giant planet, but it is tiny on the solar system's scale. If the Moon was formed far away from Earth, it must have had a very high momentum while passing near Earth, making it hard to capture. In this scenario, either the Moon would have escaped Earth's gravity or impacted Earth with a huge collision. It is also possible that the Moon could have altered Earth's orbit, causing both to fall into the Sun. Computer simulations show that such a capture would be physically impossible due to the Moon's high mass.

Daughter or Fission Theory

This theory suggests that the Moon is the daughter of Earth and that it originated from Earth itself. As we know, Earth used to spin at an immense speed during its formation. This speed slowed down over time, but it is still spinning at 1,000 miles per hour at the equator. According to the Fission Theory, a large portion of Earth was expelled due to the spinning motion. That expelled matter did not leave Earth's orbit and fall into the Sun. Instead, it started orbiting our planet. As time passed, that orbiting matter collided and clumped together, becoming bigger and bigger. After millions of years, all that matter came together and formed our Moon.

The Flaw in the Fission Theory

The Fission Theory has the same major flaw as the Sister Theory. Both Earth and the Moon have different densities. The Moon might have the same composition of some essential ingredients as we see on Earth, but the overall density is not the same.

Impact Theory

The Impact Theory suggests that the Moon was formed by the collision of a Mars-sized planet named Theia with Earth. Theia was about the size of Mars, with a diameter of about 6,102 km (3,792 miles). According to the Impact Theory, at the beginning of the solar system, a planet named Theia formed in our solar system. Evidence published in 2019 suggests that Theia might have formed in the outer Solar System rather than the inner Solar System. Since Theia did not have a stable orbit, it was pulled toward the Sun. While on its way to falling into the Sun, a small part of it collided with Earth. Such types of collisions were common in the early solar system.

It was not a direct face-to-face impact; both planets touched each other and continued moving on their paths. This impact sent billions of tons of matter shooting into space, while most of it stayed in orbit around Earth. This impact increased Earth's rotation—a day took just six hours to complete. The collision resulted in the formation of a magma belt around Earth. As time passed, the belt assembled, resulting in the formation of two similar-sized moons. However, there was a problem. These moons did not have the same velocity or orbit, which meant one more impact. Both moons collided, resulting in a single Moon orbiting Earth.

Evidence of Modern Impact Theory

In the beginning, Earth was a super-hot ball of magma. As a result, most of its heavy metals moved to the core. Theia collided with the outer part of Earth, sweeping matter from the outer layers. The primary physical evidence of this theory was seen in rocks brought back to Earth by the Apollo astronauts. On examination, it was found that the rock contained a tiny amount of iron, similar to that of Earth's outer layers. These samples also showed that the Moon's surface was once molten. Over time, it cooled down and became solid rock.

The Next 100 Years

The next 100 years are going to be the most crucial years for humanity. In the coming 100 years, what we do will decide whether humanity survives in the universe. Frank Drake was an American astrophysicist and astronomer involved in the search for extraterrestrial intelligence and alien life. He developed a mathematical equation to predict the possibility of finding civilizations in the universe. This equation considers the average life of a galaxy and stars with planets capable of developing life.

According to Frank Drake, 10,000 years is the estimated lifetime of any technological civilization. We are a technological civilization with multiple threats hovering over our heads. If we make it through the next 100 years, the possibility of making it through the next 10,000 years will increase dramatically. We are developing at an exponential pace. Just imagine where the world was 50 years ago. Remember those massive computers that no one could afford and the oversized mobile phones made for the rich only? In the last 50 years, we have taken a significant leap in technology and science. Our smartphones are a thousand times faster and more capable than the computers used by NASA to put a man on the Moon in 1969.

With the help of science, we have shaped our giant computers into small laptops. Our large 50-kilogram TVs have become slim OLED screens. The large telephones can now fit in our pockets. What technological advancements will we see in 2100? How will normal life function with the use of technology? Let us have a closer look:

1. **Future of Artificial Intelligence**: There is no doubt that Artificial Intelligence is the future. The ability of AI to learn any task by doing it repeatedly makes it unique. AI will open many doors for us and take over control of things we struggle with in our daily lives. The ability of ordinary matter to acquire such intelligence without consciousness is something beyond this world, but we have already made it happen. Today, AI can manage a home; Mark Zuckerberg spent one year developing an AI capable of handling many things at his home. It can turn off the lights, play music, entertain the kids, and so on. It cannot assemble its parts into an Iron Man suit as Mr. Stark did in the movie; it is limited to screens only. In the future, AI will be smarter—you will not need to pay your chef and gardener their salaries. AI will handle that. You will not have

to drive your car because AI will take complete control and drop you at your destination.

Some people have anxiety that AI will take over their jobs and they will have no work to do. This is a grave and genuine concern. However, as technology continues to advance, our means of generating revenue will also change. Today, more and more people are working while sitting on a chair than those doing physical work. Maybe the stock market and crypto will become a new currency, and we will not have to do much.

Others fear that AI could take over humans and rule this planet. Well, I do not think humans are dumb enough to build a machine that has the ability to take over their lives. We want to develop it, but at the same time, we want to ensure it does not cause any harm—just like what we did with the internet. However, AI can be used by bad actors. The harmful use of AI would be by people or governments using it against each other. But proper regulation could solve those problems as well.

2. **Control of Mind Over Matter:** No matter how advanced Artificial Intelligence becomes, it will always have some limitations; after all, it is artificial. Due to the evolution of millions of years, the human brain is beyond those limits. Our intelligence surpasses the limits of Artificial Intelligence. Once

we are done with AI, we will start looking at the possibilities of the human brain gaining control over physical matter. Using small chips implanted in the brain, we may be able to gain control of things around us.

The brain of an average human weighs around 3 pounds and contains more than 100 billion neurons. Neurons are cells that carry information from one place to another. Having a cell phone in our pockets enhances our capability to a large extent. We can reach out to anyone, anywhere in the world. Imagine what wonders humans could do with enhanced brain capabilities. By implanting a chip in your brain, you would be able to do most of your tasks by just thinking about them.

If you want to go somewhere, you tell the driver your destination. When AI comes, it will drive your car, but you still have to tell it where to go by inputting the destination. However, by implanting a chip in your head, you would not need to tell anyone anything. Just sit in the car and think about giving instructions to your car, and it will take you there. The same chip will control not only the car but also most of the things in your daily life. Your TV will start, your pizza will be ordered with your location, and you will be able to shop for any item from any store just by thinking about it. There would be no need to carry a six-inch smartphone.

3. **Space Travel:** Space travel opens the doors to a new future. Over time, the speed of space exploration will be boosted. I am not sure if we will be able to use wormholes or warp drives to travel vast distances by 2100—we can only hope that it will be possible in the far future. A thousand years from now, future space shuttles will use energy from stars to power themselves. These giant ships would then take us anywhere in the solar system at a speed close to that of light. We would be able to harness the energy of space to travel vast distances. The future spaceships would also run on the power of an atom; they will use nuclear reactors to power themselves. Our current thrusters will become old enough to be placed in a museum.

4. **Computers:** In the late 20th century, having a cell phone used to be a sign of wealth and well-being. When Steve Jobs announced an iPod with "1,000 songs in your pocket" in 2001, people went crazy. They had seen nothing like that. In less than 20 years, everyone is holding a smartphone with access to unlimited songs through the internet. That is how rapidly the world is changing. In the beginning, computers used to weigh multiple tons. It took an entire team to operate them. Doing some small calculations that we can now do with our calculators was the best use of computers. Today, all those computers can be found in museums.

I am sure that by 2100, today's laptops and computers will become artifacts in museums. You will no longer need to carry a smartphone or a 2–3 kg laptop in your backpack. Using the chip implanted in your brain, you would be able to turn any piece of paper into a computer and do the work you usually do. That chip will also store data for future use. The difference between a smartphone and a computer will disappear entirely.

5. **Civilization**: By 2100, we can hope to transform into a Type-1 civilization—a planetary civilization with no boundaries. We will talk about it later.

6. **Colonies on the Moon**: By 2100, we will have colonies on the Moon. Our colonies will include multiple dome-like structures. These will protect the astronauts and engineers from solar radiation and provide them with a comfortable atmosphere. The Moon has no atmosphere, and it is very harsh. The things we need to take care of are food, water, air, and solar radiation. Colonies on the Moon could become a major tourist attraction. People from different countries would be able to sign up for a tour of the Moon. A rocket from Earth will take them directly there. In general, there will be two types of tourists on the Moon—those who will land on the surface and explore, and those who will fly by the Moon and return to

Earth. This will become the new normal. An entire tourism industry for Earth's orbit and the Moon will be established.

What could be the use of a Moonbase? A base on the Moon can be used for many things. First of all, we can build a giant telescope on the Moon's far side and observe the universe at our will. NASA is already planning to do so, but no physical work has been done so far. Our Moonbase will also serve as the base for travel to Mars. When rockets take off from Earth, they burn most of their energy trying to get out of Earth's atmosphere. If we have a base on the Moon, we would not have to waste so much energy. The escape velocity of the Moon is significantly lower. If we want to land on the surface of Mars and be able to return home, we need rockets with lots of fuel. Thus, the Moonbase will serve as a good launching point.

Hubble Space Telescope

Humanity has always looked up to the stars and wondered about its place in the universe. In the beginning, we were dependent upon our eyes. All we knew came from observations made via the naked eye. Once we started looking deeper into the universe with telescopes, having a telescope on the ground was not enough. Ground telescopes can observe nearby stars and galaxies, but they tend to produce blurry images. They are not helpful for looking at distant objects. As light from distant stars and galaxies enters our atmosphere, it gets distorted by variations in temperature and density within our environment.

The presence of vast amounts of dust and other impurities further lowers the quality of the image. We see the stars twinkling at night because their light has to pass through the atmosphere before reaching our eyes. If you go outside the atmosphere and look at those same stars, they won't be twinkling—they would appear brighter and shinier. Ground-based telescopes are not fully efficient. Another problem ground-based telescopes encounter is that the atmosphere blocks or absorbs specific wavelengths of radiation, like ultraviolet, gamma, and X-rays, before they reach the ground.

To solve this problem, the idea of a telescope in space orbiting Earth was put forward by Hermann Oberth in 1923. By being outside Earth's environment, this telescope would be able to get better images and collect other scientific data. The launch of the Hubble Telescope marked the beginning of a new era in space exploration. Hubble was one of the first significant scientific instruments placed in space, launched in 1990. Scientists have used Hubble to observe some of the universe's most distant planets, stars, and galaxies. Scientists have also used the Hubble Telescope to study our solar system, not just the deep universe.

Hubble has a length of roughly 13.25 meters (43.5 feet), with a maximum diameter of 4.2 meters (14 feet). Hubble has many scientific instruments on board, which increase its weight to more than 24,000 pounds (10,886 kilograms). It orbits at an altitude of roughly 547 kilometers, at a high speed of 27,300 kilometers per hour, completing one orbit in only 95 minutes. Hubble uses its primary mirror to take most of the images, which has a diameter of roughly 2.4 meters (94.5 inches).

Hubble has two large solar panels that extend to 25 feet in direct sunlight and power six nickel-hydrogen (NiH) batteries—the powerhouse of this telescope. It has been three decades since Hubble began its work in space. Since then, more than 1.3 million different observations have taken place, based on which scientists have published thousands of research papers. Hubble's first image was taken on May 20th, 1990; it was a star cluster called NGC 3532. There are multiple instruments placed on this telescope, which act as the eyes and heart of this machine. From time to time, astronauts have gone up and replaced the older instruments. They have also installed some new instruments for better observations.

Here are some of the important scientific instruments present on Hubble:

1. **Near-Infrared Camera and Multi-Object Spectrometer:** The Near Infrared Camera and Multi-Object Spectrometer (NICMOS) is Hubble's thermal sensor. These sensors are highly sensitive to the infrared light produced by the vibrations of atoms and molecules. Many objects are hidden in interstellar dust, and NICMOS enables us to see them. One of the primary examples of such objects is stellar birth sites. When new stars are born, they are usually hidden in clouds of dust and gas. NICMOS measures the heat from that star, and based on that data, scientists can further calculate its geometry.

2. **Advanced Camera for Surveys;** The Advanced Camera for Surveys (ACS) can detect visible light. The excitation of electrons in atoms produces visible light. By measuring this visible light, we can obtain most of the information about an object. By looking deep into space, scientists study some of the earliest activities of the universe. Using ACS, scientists have measured some of the most distant objects in the universe. ACS also helps us map out the distribution of dark matter across the universe by observing gravitational lensing. The search for small or large planets around stars is carried out using ACS.

3. **Wide Field Camera:** The Wide Field Camera (WFC) can detect the spectrum of three different kinds of light: near-ultraviolet, visible, and near-infrared. It is one of the most technologically advanced instruments on Hubble. The WFC is also used to study dark energy and dark matter. This instrument generally observes galaxies that are beyond the normal observational range of Hubble.

4. **Cosmic Origins Spectrograph:** The Cosmic Origins Spectrograph (COS) acts as a prism. COS works by separating the light from different objects into its individual components. By separating the light, scientists can measure an object's temperature, density, and chemical composition.

5. **Space Telescope Imaging Spectrograph:** The Space Telescope Imaging Spectrograph (STIS) is a spectrograph that detects ultraviolet, visible, and near-infrared light. STIS is generally used to observe the larger objects in the universe. These larger objects include black holes, massive stars, and clusters.

7. **Fine Guidance Sensor:** Fine Guidance Sensors (FGS) are devices that help Hubble maintain its orientation. They help in pointing Hubble in the right direction. Hubble must be accurately aligned when observing sudden phenomena in the

universe. These devices can also measure the distance between stars and their relative motions.

All of these instruments are powered by sunlight. Hubble is useless if it is not powered by the Sun. Hubble uses large solar panels that convert sunlight directly into electricity. When Hubble is in Earth's shadow during its orbit, batteries keep it running. Initially, scientists wanted to use a nuclear reactor to power Hubble. But a nuclear reactor is not only costly but also very risky. So, the idea was abandoned.

For the last 30 years, Hubble has been working day in and day out, guiding our way into the universe. However, this machine also has its deadline. Hubble could last until 2030–2040. After this, it will retire and most likely burn up in the atmosphere. Its successor, the James Webb Space Telescope (JWST), was launched by NASA in December 2021. With improved sensitivity and resolution, JWST will be able to see what Hubble could not.

Types of Civilization

From being a single-cell organism to the most sophisticated biological beings on this planet, the creation and evolution of life have been an extraordinary journey. From what we have learned so far, evolution never ends; even today, we are evolving. We do not notice it because evolution is slow and gradual. Only after billions of years of evolution can we proudly call ourselves the most intelligent species on this planet. If evolution had stopped 6 million years ago, we would still be ape-like animals living in the forest. Not just humans—even the universe is evolving. We can measure this in terms of entropy, the extent of increasing randomness.

To a vast extent, technological advancements have taken over evolution. In the last 4 billion years, evolution has been our primary driving factor for survival—but not anymore. Where the next thousand or perhaps millions of years will lead us will be purely decided by the kind of technological advancements we make.

When it comes to our planet, we do not have complete control over it. We have to dig massive mines to extract coal and drill huge wells to take out oil buried for millions of years. Floods, volcanic eruptions, and hurricanes kill thousands of people every year, and we are still helpless.

When it comes to the solar system, we have not even reached our closest planet. We stepped on the Moon, but that was over 50 years ago. It seems like we forgot to return. Undoubtedly, Voyager 1 has crossed the solar system and reached interstellar space, but that small accomplishment alone took us over 40 years.

When it comes to the Milky Way galaxy, we find ourselves on a distant edge, orbiting a medium-sized star. In the universe, we do not even know where we stand. The only thing we know is that we don't know.

Nikolai Semyonovich Kardashev was a Soviet and Russian astrophysicist. In 1964, he proposed a scale known as the Kardashev Scale to classify the different types of civilizations living in our galaxy and the universe. This scale was created to distinguish between civilizations based on the amount and form of energy they can use. How we control different things, how we use our resources for a better future, and how far we can reach today are the parameters that will define our position on this scale. Let us look at the different types of civilizations—and where we stand—on this scale.

Type-Zero Civilization

We can pat ourselves on the back and say how advanced we are. But on the Kardashev Scale, we are just a Type-Zero civilization. A Type-Zero civilization refers to how life preserves itself under the dramatic conditions of its planet—how life sails through floods, hurricanes, and various natural disasters and still comes out unharmed. We will remain a Type-Zero civilization until we can control all these natural calamities. Since the beginning of time, the entire human history can be marked as Type-Zero because we are still struggling to fight with nature to preserve ourselves.

When life was in the water, it had its challenges—such as small creatures getting eaten by bigger creatures. For bigger creatures, it was necessary to have a particular diet every day. When life came out of the water, new challenges came to light, as it was living in a completely different environment. Today, we drive our cars on the road, but we rarely realize that we are using the energy of dead plants and animals. The fact that coal remains our primary source of energy shows how backward we still are. We are not even an intercontinental civilization because we have divided ourselves into small countries and follow different laws. The division of humanity in the form of countries protected by borders has its advantages and disadvantages. One of the advantages is that the growth of a country is directly linked to the growth of its people—but it also limits us. This division prevents us from being a planetary civilization working together for all of humanity.

The introduction of race, caste, and religion had its advantages 1,000 years ago. Race, caste, and religion brought people together in the form of small groups. They were able to relate to one another through something common. However, these things now prevent us from taking a step forward in the modern world. Whenever there is a significant scientific discovery or breakthrough, religious people often say it was already written in their book. This shows how eager we are to protect our limited religious identity.

We might have differences—religious or otherwise—but in the last 100 years, we have also taken some bold steps toward becoming a Type-1 civilization. The creation of the European Union in 1993 is just one example. It is not that humans do not want to come together and work as one. We have proven that we can act as one, but it usually happens only in extreme situations, when there is no other way or when an external threat is looming. In the 1980s, we learned that the ozone layer was depleting due to the excessive use of chlorofluorocarbons. The entire world came together and signed the Montreal Protocol to limit their use. Today, the ozone hole is healing slowly and is expected to fully recover in the next 50 years.

Generally, humans follow the tried and tested path and hesitate to do something different. However, if we want to move toward becoming a Type-1 civilization, radical changes need to happen. There is a clear indication that we're moving in the right direction. Some of those indicators include the use of the internet and the wide acceptance of English as an international language. We are living in a world that is evolving rapidly. If you went back 100 years, you might not believe how far we've come. In the last 100 years, we have gone all the way from slow bullock carts to fighter jets that challenge the speed of sound. This is a clear sign that we are transforming from a Type-Zero to a Type-1 civilization—even though we are still decades away from actually achieving it. Today is one of the most decisive times for humanity as a whole. If we make it to Type-1 civilization, that would be our biggest achievement so far.

Type-1 Civilization

We have already discussed a Type-Zero civilization and how we are slowly transitioning into a Type-1 civilization. So, what is a Type-1 civilization? A Type-1 civilization is a planetary civilization. This civilization can use and store all the energy available on its planet. In simple terms, this type of civilization can control everything happening on the planet.

A planetary civilization has the power of an entire planet in its hands. It can control earthquakes. It can use the energy bursting from a volcano. It has the technological power to stop hurricanes from causing massive damage to human lives. Instead of burning coal and petroleum and destroying nature, a Type-1 civilization can harness the energy that falls on a planet from its parent star. It can collect this energy and store it to meet the increasing demands of the population. This ability further revolutionizes its industries and technological era. This civilization will have a robust defense system in space that can deflect any asteroid that might potentially harm the lives of its inhabitants.

There will be large-scale use of nuclear power. A Type-1 civilization will use the energy of atoms, through fusion or fission, to power its industries. Harnessing Earth's energy would also mean control over the natural forces. We would be able to use wind energy, construct more dams on flowing rivers, and harness the energy of ocean waves.

How would a Type-1 civilization communicate? It will likely speak a common language. For humans, English may become the first or preferred second language. This type of civilization will be open to new ideas. Religion, as we know it, will no longer exist—or at least, most people will not consider themselves religious. Instead, they will consider themselves seekers, trying to understand their existence.

This civilization will no longer have an identity limited to caste, religion, or nation. Their identity will be cosmic—or at least human. Borders might still exist between countries, but they will hold little to no significance. There would be free movement among countries, like in the European Union. The internet is already a planetary communication system. We can contact anyone from anywhere, except for a few countries that prefer to remain secretive. Similarly, there will be a planetary flow of knowledge and culture without restrictions.

By the time we become a planetary civilization, we will have several human missions to Mars and multiple colonies on the Moon. It is not clear if we will be able to terraform and colonize Mars, but colonizing the Moon would be a great start. A planetary civilization would have complete knowledge of the Solar System. We would know if alien life exists underwater on Europa, Titan, or Enceladus. Our probes will have reached all the planets and their moons in the Solar System. Since there is a limited supply of minerals on Earth, we will begin mining asteroids for metals and other essential ores.

We are, without a doubt, progressing—but we are still another 100 to 200 years away. Famous American astronomer and science communicator Carl Sagan believed that we are currently at 0.7 on our way to becoming a Type-1 civilization.

Type-2 Civilization

After becoming a Type-1 civilization, we will most likely leave Earth. We will look for other energy sources from other parts of the solar system. A Type-1 planetary civilization can harness the energy of a planet, but a Type-2 civilization would require energy directly from its parent star. The energy of a planet is not enough to fulfill their needs, so they will look up to their star and use its energy to power their gigantic machines. This civilization is also called a stellar civilization.

To harness the energy of their star, a Type-2 civilization can use the concept of a Dyson Sphere. A Dyson Sphere is a theoretical device that would encompass a star and gather its energy. This energy would be transferred to a planet and stored for later use. A Type-2 civilization would use this energy to power their machines and spaceships that can travel to nearby stars. Not only the star, but this civilization would also utilize the hydrogen from nearby gas giants. They would drain their energy using orbiting reactors and transport that energy back to their home planet—or planets—for later use.

A Type-2 civilization would have transformed and colonized Mars. It would have colonized almost every place in the solar system where colonization is possible. A Type-1 civilization could deflect large asteroids to prevent them from hitting their home planet, whereas a Type-2 civilization could vaporize them long before they reach the planet. A Type-2 civilization would be able to move a planet from its orbit, ensuring its survival. This civilization would also send multiple rockets to explore life further into the galaxy. They would have complete knowledge of their solar system and would be ready to operate on a galactic scale. Life as a Type-2 civilization would revolve around technology and science—not family and friends.

People often ask, if there are millions of habitable planets in our galaxy alone, why don't we see aliens here on Earth? Why don't they come and visit us? Well, maybe we are not that interesting to them. Maybe they know we are here but choose to let us live comfortably and not disturb our lives. They may be choosing to ignore us because they want us to figure everything out ourselves. Or maybe they are hoping we will find them. It's also possible that they are not advanced enough. Maybe they are still fish in the ocean, evolving their way out.

Scientists predict that if we ever encounter alien life—or if aliens ever visit us—it will be a Type-2 or Type-3 civilization. This is because only such civilizations would have the capability to travel vast distances between solar systems. However, we see no evidence of alien life visiting us. We are still trying to find even a Type-Zero civilization in the solar system—perhaps on the moons of different planets.

When you become a Type-2 civilization, you are nearly immortal. There are no known forces in the universe that can easily destroy your existence. Even if your home planet is destroyed by accident, this civilization can fly to another planet in the solar system. If a supernova explosion occurs nearby, wiping out almost all forms of life, this civilization could enter its giant spaceships and move to other stars where survival is possible.

As a Type-Zero civilization transitioning into a Type-1 civilization, we are creating small probes and sending them to nearby planets and moons within the solar system. A Type-2 civilization would be able to create an army of bionic robots and send them to nearby stars so they could begin life anew. These bionic robots would transform themselves according to the planetary conditions. The environment would not be a major concern unless it is too hot or too cold. A Type-2 civilization represents a significant leap in capability and intelligence.

How close are we to becoming one? It will probably take 1,000 to 2,000 years to reach that point.

Type-3 Civilization

A Type-3 civilization is a galactic civilization. It is a civilization that can harness the energy of its galaxy. In simple terms, a Type-3 civilization would have access to the energy of an entire galaxy. If humans ever become a Type-3 civilization, our humanity would likely be left behind in the process. We would become cyborgs—beings with both biological and robotic abilities.

A Type-3 civilization would use all the tools and methods it learned as a Type-2 civilization and apply them on a galactic scale. It would be able to build Dyson Spheres all over the galaxy and harness the power of as many stars as it wants. This energy would allow them to do things that today appear only in science fiction—such as traveling at the speed of light or even through time. The civilization would not be centered on one planet or even on a few planets. Instead, it would spread across the galaxy, inhabiting as many planets as possible.

Black holes would become just another source of energy for them. They would be able to withstand the intense gravitational pull of black holes. By harnessing the massive energy from black holes, a Type-3 civilization could create wormholes and travel quickly within the galaxy—or even between galaxies.

For this civilization, there would be no boundaries known to science. They would have the ability to prevent supernova explosions. They would harness the strong magnetic fields of neutron stars for their benefit. A Type-3 civilization would not be limited to just one galaxy. As soon as a Type-2 civilization becomes capable, it would begin reaching out to other solar systems. In the same way, a Type-3 civilization would aim to reach other galaxies.

Once this civilization becomes intelligent enough, it would no longer need to steal energy from stars and black holes. Empty space is not truly empty—it holds the potential of dark matter, dark energy, radiation, and particles that constantly pop in and out of existence. To boost their rockets, they would use the energy of space itself. Because there are vast distances between galaxies, using space's ambient energy would allow them to travel between galaxies in case wormholes do not work. By doing all of this, the civilization would manipulate the laws of nature to their fullest—something we only dream of doing today.

Not just regular black holes—they would draw energy even from the supermassive black holes that exist at the center of almost every galaxy. Humans fear gamma-ray bursts, as they can be extremely harmful. For this civilization, however, a gamma-ray burst would be a source of pure energy coming directly from a black hole.

A Type-3 civilization would know everything needed to become the masters of space and time. Dark matter would no longer be a mystery, and they would be able to harness the power of dark energy to their benefit. This civilization would create colonies of cyborgs capable of self-replication and send them to nearby galaxies for exploration. Their population might also grow rapidly, as every cyborg could self-replicate and colonize every star it encounters.

A Type-3 civilization would have the highest order of adaptation. We are nowhere close to that. It would take 100,000 years or more for us to reach that level—if we ever do.

Type-4 Civilization and more

A Type-4 civilization is a universal civilization that can control all the laws of the universe. Kardashev believed that no civilization could become a Type-4 because the capabilities and power it would possess are almost god-like. Kardashev also believed that humans—or any alien species—would not be able to cope with such powers. However, some scientists believe that it is possible for such a civilization to exist, so Types 4, 5, and 6 civilizations have already been proposed.

One of the main reasons it is almost impossible for a civilization to become Type-4 is that our universe has a limited lifespan and will die one day. We have limited time to transform ourselves into beings of higher capabilities. This civilization would not abide by any known rules or laws of the universe. A Type-4 civilization would harness the energy of the universe itself. It could open the singularity of black holes and even live inside a supermassive black hole.

A Type-3 civilization would have exploited almost all known energy sources, so a Type-4 civilization would need to tap into previously unknown energy sources. They would have to discover or generate new laws of physics and govern the universe however they wished. A Type-4 civilization would be able to teleport itself through various means; wormholes are just one of them. They could create wormholes in this universe that open into another universe. That way, they would travel between universes and discover entirely new laws of physics.

Our known scientific variables cannot describe this type of civilization; it is beyond everything we currently understand. Their mental abilities would be far beyond our comprehension. This civilization would have unlocked all the mysteries of the universe. Today, we know about three dimensions of space and one dimension of time. However, theories suggest that there could be many more dimensions beyond our reach. A Type-4 civilization would be able to travel to higher dimensions—even to dimensions as small as an atom.

A Type-5 civilization can do in the multiverse what a Type-4 civilization can do in this universe. They could even create entire universes with their own unique laws and parameters. What a Type-6 civilization would be capable of, we cannot even imagine.

Humans are nowhere near reaching something like this. The first obstacle in our path is transitioning from a Type-Zero civilization to a Type-1 civilization. That would be a good start—if we don't destroy ourselves first with the various kinds of weapons we have created.

End of Life on Earth

In the last 500 million years, there have been several instances where life on this planet was wiped out. In some mass extinctions, up to 95% of all species were eliminated. One thing those extinctions had in common was that they were all natural.

Today, humanity has taken charge of this planet. A large part of what happens on Earth is now under human control. As a result, the chances of us annihilating ourselves have also gone up.

Let us look at five realistic natural and artificial ways in which this could happen.

Nuclear Warfare

Albert Einstein once said, "Two things are infinite: the universe and human stupidity, and I am not so sure about the universe." Humans might be the brainiest species on Earth, but they do not always know how to handle power with responsibility. Splitting the atom and discovering a whole new world inside it was a tremendous success, but in the end, we built nuclear bombs—bombs that can threaten humanity if placed in the wrong hands.

There was a nuclear arms race between the Soviet Union, the United States, and their respective allies during the Cold War. By the 1980s, there were over 70,000 nuclear weapons— enough to destroy this planet several times over. As the Cold War ended, both countries agreed to reduce their stockpiles significantly. I consider this one of the best decisions of the 20th century. Even though neither country agreed to reduce their stockpile to zero, the numbers fell below 14,000.

Some people might not agree, but nuclear weapons have played a significant role in maintaining global peace. We are unlikely to see another world war because of their presence. Nuclear weapons are often used as a tool for deterrence, threatening other nations and making them align with your terms. Powerful nations have been using this tactic for decades. Today, nine countries in the world possess roughly 13,900 nuclear weapons. The United States and Russia account for 91 percent of them. Many countries have given up their deadly weapons and nuclear programs in the past 30 years, whereas others have tried to acquire them.

Nuclear warfare is one of the biggest and most realistic threats that could wipe out humanity. Nuclear warfare, also called thermonuclear warfare or atomic warfare, refers to the use of nuclear weapons to destroy and damage the enemy in order to resolve political conflict. In contrast to conventional warfare, nuclear warfare is far more destructive and devastating. A nuclear weapon can release a massive amount of energy in a short period, causing long-lasting impacts on humanity, the environment, the soil, and nearly everything else.

On July 16th, 1945, the United States tested its first nuclear bomb in New Mexico. After this test, the world changed entirely within three weeks. On August 6th, 1945, the United States dropped this weapon on Hiroshima, a city in Japan. It wounded many and killed approximately 130,000 people. Three days later, another city in Japan—Nagasaki—was bombed. This blast instantly killed 74,000 people. After these two explosions, there was chaos, fear, and terror across the entire planet. These tremendously powerful blasts also ended World War II.

What would happen if global nuclear warfare occurred? When a nuclear blast happens close to the surface, soil particles mix with the highly radioactive fission products of the bomb. Some debris is transferred from the detonation site to other places by wind and falls back to Earth. A nuclear bomb causes instant casualties, injuries, and infrastructure damage from the blast and the intense heat of detonation. It is similar to a conventional bomb but on an unimaginable scale. However, it also causes long-term effects, including nuclear radiation from the initial explosion and the radioactive fallout that settles later on.

People exposed to this radiation often face genetic damage and may develop cancer. The survivors of the nuclear detonation and populations in contaminated areas are at high risk of such effects. Not just humans, but all animals are also affected by this radiation. In regions with high levels of radiation—such as old nuclear test sites—no life can survive. If it occurs underwater, no aquatic life can survive either.

The thought of nuclear war may conjure images of mushroom clouds, duck-and-cover drills, or local radiation fallout. These immediate effects are terrifying, but experts say the after-effects are even more devastating than the initial explosion.

Nuclear explosions can also eject large quantities of soot into the atmosphere, blocking sunlight from reaching Earth's surface. This phenomenon is called nuclear winter. Suppose only a small percentage of Russian and American nuclear weapons were used—it could cover the entire planet with dark clouds. In the absence of sunlight, crops would not grow. So, even if a nation survives the nuclear explosions and radiation, it would almost certainly die from hunger.

Climate Change

Another threat facing humanity is the slowly changing climate. In simple terms, climate change is the ongoing alteration in conditions such as wind, rainfall, temperature, humidity, atmospheric pressure, and other similar elements. When the usual conditions show drastic changes over a period of time, climate change is occurring. Climate change can result from both natural and human activities. Earth has experienced various dramatic climate shifts throughout its history, including the last Ice Age. Climate change caused by human activity is the result of population explosion, deforestation, excessive use of fossil fuels, automobiles, industrial waste, and other environmental hazards.

Climate change has a long history, and it is not new. We are hearing more about it today because it is becoming increasingly irreversible with every passing year. Swedish scientist Svante Arrhenius first reported climate change in 1896. He predicted that increasing carbon dioxide levels could alter the surface temperature of the Earth. The same was observed in the 1930s in the form of the greenhouse effect.

Today, these effects are becoming more apparent. Heatwaves are becoming common in Europe, Asia, South America, and Australia. Every year, we see a reduction in the amount of ice present at the North and South Poles.

During the summer, rising temperatures worldwide create genuine fear among people and climate activists. Some parts of the world have experienced temperatures up to six degrees higher than the average—for months. We should not forget that 2016 was recorded as the hottest year since we began recording temperatures. Scientists fear that as the climate continues to change, the chances of new infections, diseases, and outbreaks could increase significantly. Viruses are more likely to jump species and infect humans.

The rise of greenhouse gases and the increase in carbon dioxide emissions continue to worsen an already critical situation. It is expected that global temperature levels could rise by over 4°C by the end of this century. With rising temperatures, sea levels are projected to increase by 18 cm to 59 cm. Other expected changes include reduced snow cover, acidification of oceans, frequent heatwaves, more intense cyclones, and widespread flooding. The long-term effects of climate change are horrifying. Rising sea levels and floods could displace millions of people. Over the course of centuries, rising temperatures could raise sea levels by more than 7 meters. Many countries—such as China, India, Japan, and Vietnam—will face severe consequences.

We cannot stop climate change, but we can at least prevent it from getting worse. We can slow down its pace, which will give us enough time to handle the problem. We can put pressure on governments to act urgently. Almost everyone reading this book will be gone by the end of this century, but the next generation will have to face the consequences of climate change caused by our actions.

A Pandemic

The world has seen many pandemics throughout history. A pandemic is the global spread of a new disease, like the recent COVID-19 pandemic. In the 14th century, humanity faced a dreadful pandemic known as the Black Death. This pandemic is estimated to have killed between 75 and 200 million people—approximately 40% of the total population at the time.

The influenza pandemic (Spanish flu) of 1918 was one of the deadliest in modern history. It infected almost one-third of the global population and killed over 50 million people within a short interval, from January 1918 to December 1920. This pandemic affected people of all ages—young, old, sick, and even healthy individuals became infected, and over 10% of them did not survive.

The Spanish flu broke out in the U.S. Midwest. As World War I was ongoing, soldiers took the virus with them to Europe. Countries like Britain, Germany, France, and the United States kept it secret at first in order to maintain troop morale. When the disease became a pandemic, it appeared to have emerged in Spain, which was neutral during the war and did not censor its reporting. Hence, it was named the "Spanish flu." The flu was first suspected around March 1918. Throughout April and May of that year, the virus spread like wildfire in England, France, Spain, and Italy via army troops.

This pandemic occurred in two different phases—a milder form in early 1918, which mostly affected the sick and elderly. Most people who died during this first wave were already vulnerable due to poor health. The second wave of the virus began in August 1918, and it was far more deadly. It attacked the immune systems of younger adults. As a result, more people died in the second wave than in the first.

During this pandemic, people were struck with blistering fevers (104°F), malaise, nasal hemorrhaging, and pneumonia. Patients would often drown in their fluid-filled lungs, unable to breathe. The primary cause of the high fatality rate was pneumonia and other respiratory complications brought on by the flu. The World Health Organization (WHO) or a similar body did not exist during World War I, so the situation was uncontrollable. Nobody had a clear idea of how to handle the pandemic, and there was no global organization to guide countries in taking coordinated action to control its spread.

Worldwide, the Spanish flu pandemic of 1918 killed more people in a single year than the Black Death killed in a century. It killed more people in just 24 weeks than AIDS killed in 24 years. In India, it was referred to as "Bombay fever." The death toll in India was estimated to be between 10 and 20 million, making India the worst-hit country. A large percentage of the Indian population died during this pandemic. The period from 1911 to 1921 is the only time in recorded history when India's population actually declined.

Today, the world is more globalized than ever before. There is far more international travel for tourism and work than at any other point in history. This mobility makes it easier for a virus to spread quickly and efficiently. COVID-19 reached every continent within 3 to 4 months because people from infected regions brought it with them. The next time a pandemic occurs, the first step should be to halt global travel and quickly learn more about the virus to contain its spread.

Supervolcano

A volcano whose eruption releases more than 1,000 cubic kilometers of material is called a supervolcano. The eruption of a supervolcano is a thousand times more powerful than that of a typical volcano. Today, we have several supervolcanoes, such as the Yellowstone Caldera in the USA, Taupo in New Zealand, and Toba in Indonesia.

Globally, there are about 20 known supervolcanoes. Supervolcanic eruptions occur very rarely—only once every 100,000 years on average. But when they do erupt, they have a devastating impact on the atmosphere and climate. The last time such a volcano erupted was at Yellowstone Caldera, about 650,000 years ago. Today, it is a famous part of Yellowstone National Park in the United States. That eruption ejected over 1,000 cubic kilometers of lava and ash into the atmosphere— enough to bury an entire city several feet under volcanic debris.

The effect of such eruptions can be compared to holding a volleyball underwater. When you release it, the air-filled volleyball is forced upward by the high-density water. As a result, we see a quick and violent eruption.

Yellowstone Volcano

Yellowstone is one of the world's oldest national parks, established in 1872. Yellowstone covers 8,987 square kilometers and spans multiple states. Nearly 3 million people visit this park every year to enjoy its stunning natural landscape, which includes hiking trails, geysers, hot springs, and mountain peaks.

Beneath the surface of this park lurks another natural wonder—one with the power to wipe the park off the map.

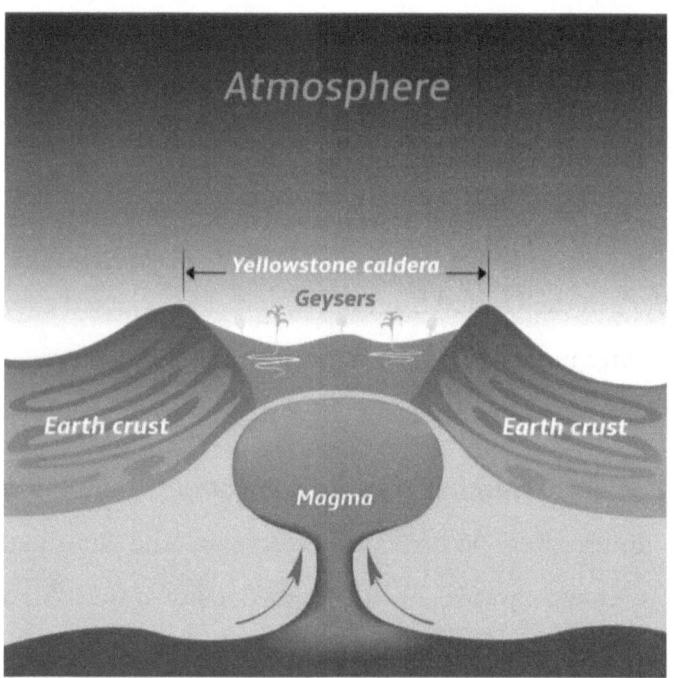

The Yellowstone Volcano is one of the largest known supervolcanoes. Its appearance is not like a typical cone-shaped volcano. The magma of the Yellowstone volcano is extremely explosive, thick, and almost paste-like. Although there is no indication of an eruption anytime soon, a full explosion would be disastrous for nature lovers and people living near the park.

If this volcano were to erupt, heat rising from within the Earth's core would begin to melt the molten rock beneath the surface. This would create a mixture of rocks, magma, vapor, carbon dioxide, and other gases. As this mixture accumulates and rises over thousands of years, the pressure would push the ground upward into a dome or hemispherical shape and create cracks along the edges. When the pressure is released through these cracks, the dissolved gases would explode, rapidly spreading magma across the park.

Its eruption could kill as many as 100,000 people instantly and spread a 3-meter layer of molten ash as far as 1,600 kilometers from the park. Rescuers would have an extremely difficult time getting into the area. The ash would block all entry points from the ground and release gases and ash into the atmosphere, disrupting most air travel—just as it did when a much smaller volcano erupted in Iceland in 2010.

The aftereffects of such an eruption are equally as frightening as a "nuclear winter." It could blanket the planet with ash and dust, drastically cooling temperatures and damaging ecosystems. The good news is that a large-scale eruption is extremely unlikely to occur in our lifetime. Yellowstone last erupted about 650,000 years ago. The United States Geological Survey says the probability of it blowing its top again anytime soon is very low.

Asteroid Impact

The word "asteroid" means "star-like." Asteroids are mineral-rich space rocks, smaller than planets, and are mainly found in the asteroid belt around the Sun. They should not be confused with comets, which are made of dust and ice. Comets remain icy as they orbit far from the Sun. If they get too close to the Sun, its intense heat will melt them.

Undoubtedly, asteroids are smaller than planets, but some are large enough to end all life on this planet. If a giant asteroid were to hit a landmass, a tremendous amount of soil and dust would be thrown into the atmosphere, resulting in the loss of animal and plant life—similar to a nuclear winter caused by nuclear explosions. However, if an asteroid hits the ocean, it could trigger tsunamis, hurricanes, and earthquakes in coastal regions due to the release of a massive amount of energy.

Large asteroid impacts have devastated the planet in the past. Dinosaurs, which once ruled this planet, are no longer here. To prevent a future asteroid impact, we first need to know the location and trajectory of the asteroid. Today, we have many observatories around the world dedicated to this purpose. They have calculated the trajectories of thousands of asteroids. We have identified almost all the giant asteroids capable of causing a global disaster. However, many smaller asteroids capable of causing regional destruction are still undiscovered.

Here's how we can prevent an asteroid from crashing into our planet.

Target Change: Changing the trajectory of an asteroid would be the easiest and safest way to prevent a disaster. Suppose there is an asteroid, 1 billion kilometers away, heading straight toward Earth. If we change its trajectory by just 1 degree, that would be enough to prevent a collision. By the time it travels 1 billion kilometers, its path will have shifted by hundreds of thousands of kilometers.

Asteroid Destruction: If changing the trajectory is no longer an option—perhaps because the asteroid is too close—then we might consider destroying it. Blowing the asteroid into small pieces would reduce its impact. Even if some fragments enter Earth's atmosphere, most of them would burn up before reaching the surface.

Most of the asteroids in our solar system orbit within the asteroid belt between Jupiter and Mars. A small percentage of them are larger than one kilometer in diameter, but the majority are much smaller. These asteroids generally pose no threat, as they are fixed in their orbits. However, Earth-crossing asteroids—those that intersect Earth's orbital path—and Near-Earth asteroids—those that come close—are more concerning. There are roughly 10,000 known Near-Earth asteroids. Of these, about 860 are larger than one kilometer, posing a significant threat. The total number of potentially hazardous asteroids is around 1,400.

Apophis is a Near-Earth asteroid with a diameter of 370 meters and an average orbital speed of 30 km/s. In December 2004, it caused a brief period of concern when initial observations indicated a 2.7% chance of it hitting Earth. That probability has since been significantly reduced. Astronomers now say that Apophis has no chance of impacting Earth in the near future.

Asteroids with diameters of around 1 kilometer hit Earth once every 500,000 years on average. Larger asteroids—those with diameters greater than 5 kilometers—collide with Earth once every 20 million years. Even though we have identified many potentially hazardous asteroids, the chances of one hitting our planet remain significantly low.

PART–IV

Death of the Universe

Different scientists predict the future of our universe in different ways—with some proposing a finite age and others suggesting an infinite one. However, not everyone can be correct. The death of the universe is one of the biggest topics of discussion among scientists. From what we have learned so far, the fate of our universe falls under the field of physical cosmology and is directly influenced by the role that dark energy and dark matter play as the universe continues to age.

There is one strict rule that all life follows—everything comes to an end. Does this rule apply to the universe as well? Perhaps yes. But when that happens, no one will be there to witness it. The fate of our universe includes not only the death of all stars, solar systems, and galaxies but also the end of all life. It will be the death of intelligence and consciousness—something that took billions of years to evolve into what it is today.

Our universe is approximately 13.8 billion years old. What the next 13.8 billion years will look like can only be imagined. Ever since Edwin Hubble pointed out that galaxies are moving apart, scientists have been trying to predict the ultimate fate of the universe. Most theories about the end of the universe are based on what Hubble discovered.

Let us now explore the different ways in which the universe could end:

Big Freeze

The Big Freeze, also known as the "future of the ever-expanding universe," is one of the most widely accepted scenarios for the death of our universe. It describes a situation in which the universe's continuous expansion results in a cosmos that approaches absolute zero temperature.

Absolute zero is the lowest possible temperature a system can reach, measured as -273.15 degrees Celsius. At this temperature, the heat energy within a system is zero and cannot decrease any further.

Various scientific observations also suggest that the universe's expansion will continue indefinitely, making the Big Freeze a likely outcome for its ultimate fate.

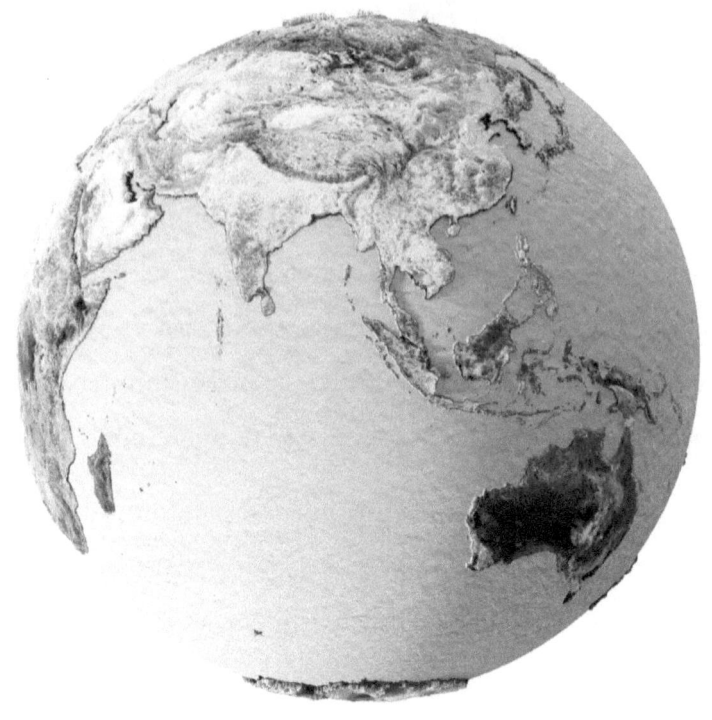

Although we are yet to fully understand what dark energy is, what we have learned so far has given rise to the idea of the Big Freeze. The universe is undergoing a one-way expansion, with dark energy as the dominant force. We cannot control this expansion—it is beyond our grasp. As a result, the universe will continue to expand, and dark energy will grow even stronger, further accelerating this expansion. With this acceleration, the universe will expand so rapidly that galaxies will break the cosmic light barrier and become invisible forever.

There is a limited amount of gas and dust in galaxies—materials necessary for star formation. In the next 1 trillion years, there won't be any more gas or dust clouds left, as galaxies will have already used them up to form stars. Once galaxies run out of gas and dust, the formation of new stars will cease. At that point, there will be a finite number of stars in the universe. As we know, stars have limited lifespans due to their limited fuel. Initially, the larger stars will exhaust their fuel, resulting in supernova explosions. After that, smaller stars will run out of fuel and become white dwarfs. At this stage, the universe will grow dim, with only a limited amount of heat and light remaining.

Stars are the powerhouses that light up the entire universe. Without them, everything will start cooling rapidly. Even the white dwarf stars will begin to cool. Eventually, they will become so cold that you could touch them without being burned. The fusion reactions at their cores will stop completely, and they will freeze. With no new source of heat or light, the universe will turn cold and dark—approaching absolute zero.

Black holes would still remain, but they are not a reliable source of heat or light. Any advanced civilizations (if they exist) might attempt to survive near black holes and use their gravitational force to generate heat and electricity. However, survival in such regions would be incredibly difficult due to the minimal availability of resources. Such civilizations may survive for a while—but not forever. No creature could endure such a cold, lifeless universe indefinitely.

Some scientists have suggested that gravity could eventually slow down or reverse the expansion of the universe. However, there is not enough matter in the universe to overpower the force of dark energy. Matter makes up just 4.9% of our observable universe. In comparison, the overall strength of gravity is too weak to counteract dark energy. Only a miracle could save the universe from this fate.

The majority of scientists believe that the Big Freeze is inevitable. No matter what we do, it will happen one day. No known force in science can stop this expansion. Research is ongoing into the nature of dark energy. Perhaps, as we learn more about its repulsive properties, we might discover a way to avoid this fate. But as of today, there is no hope for a future that does not end in absolute zero.

Big Rip

The Big Rip is an extension of what we have learned in the Big Freeze scenario for the death of the universe. It is a hypothesis in which the universe continues expanding—and does so at such a rapid rate that everything in existence will be torn apart and reduced to pure energy.

As we know, the universe is expanding at an accelerating rate. With this expansion, the strength of dark energy is also increasing, causing even faster acceleration. This theory states that the universe will continue expanding at an ever-increasing pace. The Big Freeze predicts that galaxies will move apart but remain structurally intact. However, the Big Rip suggests that even galaxies themselves will begin to stretch—and ultimately be destroyed—due to this extreme expansion.

All the stars we see in the night sky belong to our Milky Way galaxy, located within a few thousand light-years of Earth. But as the universe continues expanding, nothing will remain visible in the night sky. Eventually, even Alpha Centauri, our nearest star, will surpass the light-speed barrier and vanish from view—forever.

But that's not all. Even our solar system will be torn apart. The distance between the Sun and Earth will increase. All the planets will start moving away from one another. Even more fascinating (and terrifying) is that this expansion will occur on the atomic scale as well. If the rate of expansion becomes fast enough, atoms will begin to swell. The distance between electrons and their nuclei will increase until the electrons can no longer orbit the nucleus. The protons and neutrons that make up atomic nuclei will lose their binding force and be pulled apart. Eventually, even these particles will disintegrate into their most basic constituents and cease to interact with one another.

This gets horrifying when we realize that we are made up of atoms. When this happens, we would feel our bodies swelling—their length and width increasing. Of course, this would cause unimaginable pain. But we wouldn't survive for long. Our arteries would rupture, and most of our internal organs would fail as our atoms were pulled apart. Eventually, everything in the universe would be reduced to pure energy—just as it was at the beginning of the universe.

From the point of view of String Theory, nothing would remain in the universe except for vibrating strings—the most fundamental building blocks of matter. This idea is terrifying but seems plausible, considering the repulsive power of dark energy. While dark energy is accelerating the universe's expansion, it is not only working against the attractive force of ordinary matter, but also against dark matter. If dark energy can overcome both, it would have enough force to rip apart everything—down to the very fabric of matter.

What would happen to black holes in the Big Rip? It's fascinating to examine the fate of the universe from the perspective of black holes, since they appear nearly immortal. However, as we've discussed before, Hawking radiation slowly causes black holes to evaporate. All black holes—including the supermassive ones—emit energy in the form of Hawking radiation. As the universe continues its exponential expansion, the rate of Hawking radiation will increase. Black holes will eventually weaken and radiate away.

That said, this process will take far longer than the disintegration of ordinary matter. Once a black hole has evaporated, all that will remain is a smooth, uncurved, and endlessly expanding space.

Big Crunch

The Big Crunch is quite different from what we've discussed so far. It is essentially the opposite of the Big Freeze scenario. This theory assumes that the average density of the universe is sufficient to halt the expansion—eventually overpowering it with gravitational attraction.

According to this theory, the universe is currently expanding at an exponential rate due to the effects of the Big Bang. However, over time, this expansion will slow down. As the expansion decelerates, the attractive forces of matter and dark matter will begin to dominate, causing the universe to start contracting.

Due to this contraction, all galaxies will begin moving closer together. The diameter of the universe will shrink. All the stars in our night sky would appear brighter and brighter. The planets of our solar system would come closer and closer to each other. The Sun would appear larger and more luminous, and the Earth would receive more heat.

As galaxies, solar systems, stars, and planets continue drawing closer, the rate of contraction would increase. The universe's overall density would rise dramatically. Eventually, everything would collapse into a hot plasma ball, and finally into a dimensionless singularity. Gravity would pull everything back into a single point—just as it may have before the universe began.

According to this theory, the Big Bang could occur immediately after the Big Crunch, restarting the universe in an endless cycle. This concept supports a cyclical or oscillating universe model: a repeating sequence of formation, expansion, contraction, and rebirth.

This theory even attempts to answer the age-old question: What existed before the Big Bang?

However, everything we've discovered through observation currently goes against this theory. The evidence indicates that the universe is not a closed system, which has led cosmologists to abandon the oscillating universe model.

A Big Crunch would result in the universe's heat death, but that contradicts current data. Modern observations show that the expansion of the universe is not slowing down or remaining constant—it is accelerating. With the growing influence of dark energy, it appears unlikely that the universe will end in a Big Crunch.

Instead, the Big Freeze and Big Rip are considered the more plausible end-of-universe scenarios in today's cosmology.

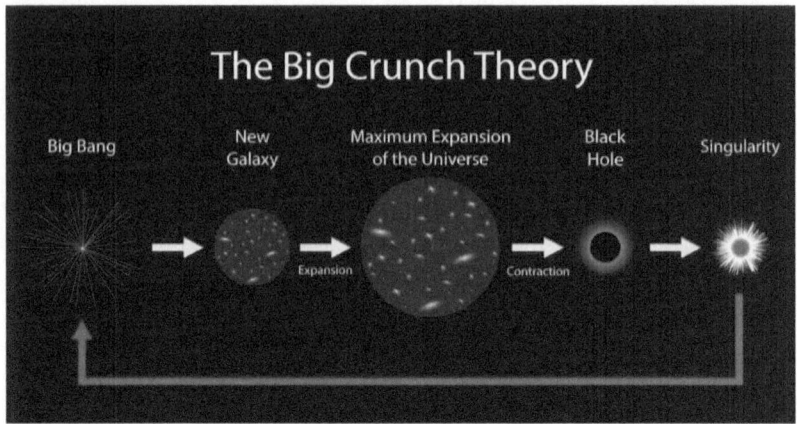

The Big Bounce is another model of the universe that strongly supports the idea of the Big Crunch. It suggests that a cyclical or oscillatory universe is possible. In this model, every Big Bang is the result of the collapse of a previous universe.

Some scientists predict that dark matter may not originate from our dimension; instead, it could be part of higher dimensions that are currently leaking into our universe. If a significant amount of dark matter were to suddenly leak into our universe, it could greatly increase the gravitational force, making the Big Bounce scenario more likely.

Apart from this theoretical possibility, there is currently no substantial hope for this particular fate of the universe, based on current observational evidence.

Other Fates

Eternal Inflation: Eternal inflation is a model of a hypothetical inflationary universe. It is an extension of the Big Bang theory. While the Big Bang explains the universe's birth, it does not account for its ultimate fate.

According to the theory of eternal inflation, our universe is like an inflating bubble. All the matter in the universe—including stars and planets—exists on the surface (or "skin") of this bubble. Since the Big Bang, this bubble has been expanding, and it will continue to expand forever. No known force in the universe can stop this expansion.

The eternal inflation theory supports the Big Freeze model, in which the universe will eventually freeze to death. Hubble's law also aligns with this view, suggesting that the future of the universe is inflationary.

However, this theory contradicts the Big Rip model of the universe and suggests that expansion could never occur at the atomic scale.

False Vacuum: Consider a balloon filled with air. The balloon is inflated by an external force and appears very stable. The air isn't leaking out, and everything seems to stay right where it is. However, that stability is only an illusion—maintained by the surface of the balloon, which holds all the air and prevents it from escaping.

If you pop the balloon, the air will rush out and settle into a state of minimum potential, equal to the surrounding atmospheric pressure.

Some scientists believe that the universe we live in might not have reached its lowest and most stable energy state yet. If we are indeed inside such a balloon, we would never see it coming. When this fate occurs, the very fabric of space-time will collapse. All the known forces of the universe will cease to function, and the universe as we know it will end.

As we know, vacuum itself has energy. Therefore, the universe might one day try to reach a state of minimum energy. While the universe has existed for billions of years and nothing catastrophic has happened yet, this leads some to believe that there may not be such a thing as a "false" or "true" vacuum.

Therefore, many argue that this scenario—while theoretically interesting—may not be a cause for concern.

Timeline of the Future

Event	Number of Years
Halley's Comet Visit	49
Colonies on the Moon	80
Antares Supernova Explosion	10,000
Niagara Falls Erodes Away	50,000
VY Canis Majoris explosion	100,000
Humans Terraform Mars	100,000
Supervolcanic Eruption on Earth	100,000
WR 104 Explodes into Supernova	300,000
Earth Likely Hit by 1 km Asteroid	500,000
Pyramids of Giza Erodes Away	1 million

Humanity Colonizes Milky Way	1 million
Likely Supervolcanic Eruption	1 million
Gliese 710 To Pass Within 9,000 AU	1.4 million
Grand Canyon Will Erode	2 million
Phobos Collision with Mars	50 million
Saturn Loses Its Rings	100 million
Sun's Luminosity Increased by 1%	110 million
One Day is 25 Hours Long	180 million
Solar System Completes 1 Galactic Year	240 million
New Supercontinent	250 million
Near Gamma-Ray Burst	500 million
CO_2 Levels Too Low for Photosynthesis	700 million

Death of All Plant Life	800 million
Earth's Oceans Starts Evaporating	1 billion
Sun's Luminosity Increased by 10%	1.1 billion
Earth's Oceans Evaporate Away	2 billion
Death of Most Life on Earth	2 billion
Sun Expands into a Red Giant	4 billion
Andromeda Collision with Milky Way	4.5 billion
Sun Destroys the Earth	7.9 billion
Sun Becomes a White Dwarf	8 billion
Moon Collision with Earth	65 billion
Universe End Via Big Crunch	1 trillion
Peak Habitability in the Universe	10 trillion

Formation of New Stars Ends	100 trillion
All Stars Exhaust Their Fuel	110–120 trillion
Universe Will Become Completely Dark	150 trillion
Sun Cools Down to -268 °C	1 quadrillion
Nucleons Start Decaying	2 undecillions
Black Hole TON 618 Dissipates	0.6 googol
Black Hole Era Ends	1,700,000 googols

Glossary

Absolute Zero: The lowest possible temperature a body can reach; its value is -273.15°C.

Anti-gravity: The opposite of gravity. Anti-gravity, often associated with dark energy, is believed to be responsible for the universe's accelerated expansion.

Antimatter: The opposite of ordinary matter. Antiprotons have a negative charge, and positrons (antielectrons) have a positive charge. Every particle is believed to have a corresponding antiparticle.

Atom: The basic unit of matter, consisting of a nucleus made up of protons and neutrons, surrounded by electrons in motion.

Big Bang: A scientific theory explaining the origin of the universe. Evidence shows that this cosmic explosion occurred approximately 13.8 billion years ago.

Big Crunch: A hypothetical fate of the universe in which gravity overcomes dark energy, causing the universe to collapse into a singularity, potentially leading to another Big Bang.

Big Freeze: A theory suggesting the universe's expansion will continue forever. Stars will burn out, black holes will evaporate, and the universe will reach absolute zero, ending all intelligent life.

Big Rip: A theory suggesting that the accelerating expansion of the universe will eventually tear apart galaxies, stars, planets, and even atoms, reducing all matter to its fundamental particles.

Black Hole: A region in space where gravity is so strong that not even light can escape. Its escape velocity exceeds the speed of light.

Blue Shift: A phenomenon caused by an object moving toward the observer. For example, a yellow star approaching at high speed appears blue due to compressed light wavelengths.

Chandrasekhar Limit: The maximum mass a white dwarf star can have (about 1.4 times the mass of the Sun) before collapsing into a black hole or neutron star.

Conservation Law: A fundamental law of physics stating that energy cannot be created or destroyed, only transformed from one form to another.

Cosmic Microwave Background (CMB): The residual thermal radiation from the Big Bang, often called the "afterglow" of the universe's birth.

Cosmological Constant: A term introduced by Einstein in his equations to represent a static universe. He later called it his "greatest blunder" after the discovery that the universe is expanding.

Cosmology: The scientific study of the origin, structure, evolution, and fate of the universe.

Critical Density: The exact average density of matter required for the universe's expansion to halt. Current data suggests the universe is not dense enough, allowing dark energy to dominate.

Dark Energy: A mysterious form of energy that makes up about 68.3% of the universe. It is responsible for the accelerating expansion of the universe.

Dark Matter: A form of invisible matter making up about 27% of the universe. Its existence is inferred through its gravitational effects on visible matter and light.

Dimension: A measurable extent of space or time. The most commonly known dimensions are length, width, height, and time.

Electromagnetic Force: One of the four fundamental forces of nature, responsible for electricity, magnetism, and light.

Electron: An elementary particle discovered by Sir J.J. Thomson. It carries a negative charge of -1.602×10^{-19} C and has a mass of 9.109×10^{-31} kg.

Entropy: A measure of randomness or disorder in a system. Gases have higher entropy than solids or liquids. In any closed system, entropy increases over time, according to the second law of thermodynamics.

Eternal Inflation: A model of a hypothetically inflationary universe. It extends the Big Bang theory, proposing that the universe is continually expanding in a bubble-like form.

Event Horizon: The boundary surrounding a black hole, beyond which nothing—not even light—can escape.

LASER: Acronym for Light Amplification by Stimulated Emission of Radiation. A laser emits a highly focused, coherent, and monochromatic beam of light.

False Vacuum: A vacuum state that has not yet reached its lowest energy level. Similar to how a supernova reaches stability through collapse, the universe could suddenly shift from a false vacuum to a true vacuum, potentially ending reality as we know it.

Frequency: The number of cycles or vibrations per second, measured in hertz (Hz).

Fusion: A process where atomic nuclei combine, forming a new element and releasing energy. This is the process that powers stars.

Galaxy: A massive, gravitationally bound system of stars, planets, gas, dust, and dark matter. Galaxies come in elliptical, spiral, or irregular shapes. Our Milky Way galaxy contains an estimated 100 to 400 billion stars.

General Relativity: A theory developed by Albert Einstein in 1916, describing gravity as the curvature of space-time caused by mass and energy.

God: Often regarded as the creator and ruler of the universe; a concept rooted in faith and belief systems, not scientific measurement.

Grand Unification Theory (GUT): A theoretical framework that attempts to unify the electromagnetic, weak nuclear, and strong nuclear forces. GUT does not yet incorporate gravity.

Gravitation: The weakest of the four fundamental forces, responsible for the attraction between masses. It keeps planets in orbit and holds us to the Earth's surface.

Higgs Boson: A fundamental particle associated with the Higgs field. It is responsible for giving other particles their mass.

Higgs Field: An energy field believed to have existed shortly after the Big Bang. It interacts with particles to give them mass.

Hubble's Law: States that the recession velocity of a galaxy is directly proportional to its distance from us, providing key evidence for the expanding universe.

Large Hadron Collider (LHC): The world's largest and most powerful particle accelerator. It collides high-energy particles (not photons, but usually protons) to study the fundamental components of matter.

Light: A form of electromagnetic radiation that travels in waves. It includes a spectrum of wavelengths from visible light to infrared and ultraviolet.

Light-Year: The distance that light travels in one year—approximately 9.46 trillion kilometers (5.88 trillion miles). Used to measure astronomical distances.

Magnetic Field: A field produced by moving electric charges, which exerts force on other moving charges and magnetic materials.

Mass: A measure of the amount of matter in an object. It also determines an object's resistance to acceleration and how it responds to gravitational forces.

Matter: Anything that has mass and occupies space. It is composed of atoms and molecules, and includes everything that can be seen or touched.

Multiverse: The concept of multiple universes. It suggests that there could be infinite universes, and we live in just one of them.

Neutron: A neutral subatomic particle with a mass of 1.672×10^{-27} kg. Along with protons, it makes up the atomic nucleus.

Neutron Star: The collapsed core of a massive star formed after a supernova explosion. Neutron stars are among the smallest and densest known stars.

Nucleus: The dense, positively charged center of an atom, composed of protons and neutrons.

Particle Accelerator: A machine that accelerates charged particles to high speeds, often used in nuclear and particle physics experiments.

Photon: An elementary particle with zero rest mass; it is the fundamental constituent of light. Photon absorption can excite electrons, enabling effects such as the photoelectric effect.

Proton: A positively charged subatomic particle with a charge of 1.602×10^{-19} C and a mass of 1.672×10^{-27} kg.

Quantum Mechanics: The branch of physics that studies phenomena at atomic and subatomic scales, focusing on low-energy interactions.

Quarks: Elementary particles that combine in groups of three to form protons and neutrons.

Radioactivity: The spontaneous decay of unstable atomic nuclei. During this process, a nucleus breaks down into daughter nuclei. The rate of decay depends on the material.

Red Shift: The phenomenon in which light from distant celestial objects shifts toward the red end of the electromagnetic spectrum, indicating that the universe is expanding.

Singularity: A point of infinite density where the known laws of physics break down. The universe is believed to have originated from a singularity, and they are also found at the centers of black holes.

Space: A three-dimensional continuum in which objects have position, direction, and movement relative to one another.

Space-time: A four-dimensional concept combining space and time into a single framework. It is fundamental to Einstein's theory of relativity.

Special Relativity: Proposed by Albert Einstein in 1905, it states that the laws of physics remain constant for all observers, regardless of their constant velocity.

Steady-State Theory: A theory suggesting that the universe has no beginning or end and has always existed. It proposes a continuous creation of matter to maintain a constant density.

String Theory: A theoretical framework proposing that all particles are made of tiny, vibrating one-dimensional strings, rather than point-like particles.

Strong Nuclear Force: The strongest of the four fundamental forces. It binds protons and neutrons together inside the atomic nucleus.

Supernova: A massive explosion that occurs at the end of a star's life cycle, releasing an enormous amount of energy and light in a short time.

Time: The continuous progression of events from the past, through the present, into the future.

Time Travel: The concept of moving backward or forward through time, often explored in science fiction and theoretical physics.

Types of Civilization: A classification system (like the Kardashev Scale) that categorizes civilizations based on their energy usage and control—from planetary to galactic levels.

Unified Field Theory: A theoretical framework, pursued by Albert Einstein, that aims to unify all four fundamental forces: gravitation, electromagnetism, strong nuclear force, and weak nuclear force.

Universal Forces: The fundamental forces that govern all interactions in the universe: gravitational, electromagnetic, strong nuclear, and weak nuclear forces.

Vacuum: Empty space that contains no matter. However, even a vacuum possesses energy—known as vacuum energy—according to quantum physics.

Virtual Particles: Short-lived particles that appear and annihilate within a fraction of a second, in accordance with quantum field theory.

Visible Light: The part of the electromagnetic spectrum detectable by the human eye, with wavelengths ranging from ~390 to 700 nanometers.

Weak Nuclear Force: A fundamental force responsible for radioactive decay and nuclear fusion in stars.

Weight: The gravitational force experienced by a body due to its mass and the local gravitational field.

White Dwarf: A stellar remnant formed when a medium-sized star exhausts its fuel. It is small, dense, and slowly cools over time.

Wormholes: Hypothetical tunnels in space-time that could act as shortcuts between distant parts of the universe—or even different universes.

X-Rays: A form of electromagnetic radiation with wavelengths between 0.01 and 10 nanometers. X-rays are used in medical imaging and astronomy.

The Purpose

We have done remarkable work in cosmology in the last 500 years. However, one question still bothers us: Does the universe have a purpose? When we look deeper into the universe, we do not seem to find a reason for its existence. Maybe it was created for a reason that is too complicated for us to understand. But whenever we talk about purpose, scientists are divided into two groups:

The first group of scientists believes that the universe has a purpose. Most of them also believe in the existence of God. The God who created us must have a plan. The universe has a purpose because life, from its very first spark, is governed by various laws of nature. All the past events through which life came out of the water and became what it is today cannot be a mere coincidence. We find a similar pattern if we look at the universe from an atom to the solar system and to the giant galaxies. There is a set of rules and laws that govern them, and we must be thankful for that. Maybe we will discover the purpose in a million years, but until then, we have to keep this cycle going.

The question "What or who is God?" is a valid question in science, so finding an answer becomes our responsibility. There was a time when it was believed that whatever was happening was a divine act of God, and it would be a sin to question it. However, not everyone believed this. Some people stood for themselves and questioned the existence of God even though they feared going straight to hell. This curiosity, even to question the existence of God, has helped us make various advancements in science. If no one had questioned God's existence, there would be no agnostics, atheists, or spiritual people.

There is a misconception that since many scientists believe in God, there must be a God. Well, scientists who believe in God do not use God or their holy books such as the Bible, Qur'an, or Gita for their scientific research. They use religion for spiritual enlightenment, finding a purpose**,** or maybe during an existential crisis. When it comes to science, religion has no significant role to play. When I see so much suffering in the world, people dying in floods while others are dying in droughts without water, it makes me question the existence of a God. Maybe God does not care about human suffering, or maybe God does not exist at all.

The second group of scientists believes that the universe does not have a purpose because we do not play a central role in the universe. We are tiny creatures living on a small planet, circling an average-sized star in an average-sized galaxy somewhere in the universe. When it comes to our role in the universe, we do not matter to the cosmos. The universe will go on its way, whether we are present or not. There are universal forces inside us that support and nourish life, but an equal number of forces are trying to kill us. Every day, people are dying due to all sorts of silly reasons.

Every passing moment, the universe is growing toward its fate. Maybe in a trillion years from now, the universe will die out. So, what is the meaning of our existence? A universe without a purpose does not mean that our lives are purposeless. The universe might not have a purpose for us, but we create our own purposes. For people who believe in religion, religion gives them a purpose to serve a God, go to heaven, and live forever.

Today, we struggle to find our place in the universe, and it is even harder to predict our purpose. The search for a purpose creates a purpose in itself. If the universe does not have a purpose, it will die anyway. That does not make our lives meaningless. We are capable of creating meaning. Life starts and ultimately ends, then it starts again[**],[**] but not from where it ended. It is hard for science to say whether we have a purpose. Science itself goes through a continuous state of change. Only religion can answer this question confidently, but we must not forget that "religion cannot show it, and science cannot prove it today."

The Conclusion

American astronaut Gene Cernan, also known as the "last man to walk on the Moon," once said, "Curiosity is the essence of our existence." If the Wright brothers had not been curious about how things could fly, we would not have airplanes. If Sir Isaac Newton had not been curious about the falling apple, he would have never discovered the laws of gravity. If the first humans had not been curious about new lands, we would still be living in Africa. Curiosity is the mother of all inventions; it propels the wheel of innovation and discovery.

Our view of the universe was different ten years ago, and I am sure it will be significantly different ten years from now. The universe will continue to make us wonder, and the day it stops doing so, we will be doing something wrong. We must look back and start fresh. Today, we know a little about the universe because we stand on the shoulders of giants such as Sir Isaac Newton and Albert Einstein, whose teachings stand right in the middle of our understanding of the universe.

As a kid, I always looked up to many scientists and thinkers who helped me understand how little I knew. Some of them were Professor Stephen Hawking, Dr. Michio Kaku, and Dr.

Neil deGrasse Tyson. Studying their work and watching them has made me realize how vast the cosmos is. A universe without us is not a universe. It is nothing because there is no one else to understand its language. The role of science is to do just that.

As you move on in life, there are two ways to live**:** either you do not know that you have never opened your eyes because of some strong faith, or you do not want to close your eyes because the beauty of the universe is endless. Above all, we must remember that science is the only way forward.

Stay curious!